W0112356

ELECTROSORPTION

Contributors to this Volume

S. D. Argade
M. A. Genshaw
E. Gileadi
H. D. Hurwitz
K. Müller
B. J. Piersma
A. K. N. Reddy

ELECTROSORPTION

Edited by
Eliezer Gileadi

Electrochemistry Laboratory
John Harrison Laboratory of Chemistry
University of Pennsylvania
Philadelphia, Pennsylvania

and

Institute of Chemistry
Tel-Aviv University
Tel-Aviv, Israel

With a Foreword by John O'M. Bockris

 PLENUM PRESS · NEW YORK · 1967

ISBN-13: 978-1-4684-1733-3   e-ISBN-13: 978-1-4684-1731-9
DOI: 10.1007/978-1-4684-1731-9

Library of Congress Catalog Card Number 67-15143

© *1967 Plenum Press*
Softcover reprint of the hardcover 1st edition 1967

A Division of Plenum Publishing Corporation
227 West 17 Street, New York, N. Y. 10011
All rights reserved

No part of this publication may be reproduced in any
form without written permission from the publisher

Foreword

The gradual emergence during the last decade of the study of the mechanism of electrode reactions from the dark ages has given stimulus to a consideration of the double layer at metal–solution interfaces, which extends far outside the classical experimental studies of the capacitance of the mercury solution interface made during the 1950's by D. C. Grahame at Amherst College, Massachusetts.

The central aspect of the study of an electrode reaction is the elucidation of its path and rate-determining step. Two fields are, however, prerequisites for such studies. First, it must be known what species are in the bulk of the solution, for these will seldom be simple ones such as H_3O^+, and this study ("complex ions") has been made with both extent and depth. Second, the occupancy of the surface of the electrocatalyst and the associated field gradients must be known as a function of position in the double layer. Such "maps of the double layer" can be given with reasonable certainty up to concentrations of about 1 N for mercury in contact with solutions of inorganic ions. However, this is—or was until very recently—the extent of the knowledge. The problems confronting a fundamental approach to the rational development of, e.g., fuel cell catalysis were therefore considerable.

A beginning has now been made on the general problem of electrosorption, and Dr. E. Gileadi has taken the initiative of bringing together the workers who were engaged in the Electrochemistry Laboratory, University of Pennsylvania in 1965–66 on double layer studies and getting them to give a course of lectures that attempts to present the position at this time.

Technologies are of two types: (1) those which developed in parallel to, or behind, the understanding of what went on in the processes concerned (electronics and atomic energy) and (2) those in which the technology was pulled far ahead of the understanding and aided by a good dose of Edisonianism. The vast electrochemical industries (nonferrous metallurgy; electrosynthesis; energy storage; electroplating; and, recently, energy conversion) certainly await the injection of a considerable amount of understanding in order that they may move away from the unchanged positions many have had for decades. Molecular biology demands, in many systems, electrochemical mechanisms, and it, too, may be aided by an increased understanding of electrosorption. The control of the stability of materials is not a separate technology, but in every technology it depends on a knowledge of the structure of the interfacial region and is specifically controlled by electrosorption.

All these parts of applied chemistry depend for progress on the concepts and mechanism elucidation made in fundamental chemistry, and, particularly in the tremendous lacuna of modern chemistry, progress in the fundamentals of charged interfaces. On one important side of this, Dr. Gileadi's book should be a great help to all who are making interfacial studies and particularly to the technologists in the above areas who are trying to make progress with devices, often with insufficient or negligible understanding of the structure of the interface.

John O'M. Bockris

Philadelphia, Pennsylvania
March 1967

Preface

The idea of producing a volume on electrosorption came up following a seminar series on this subject given by senior members of the Electrochemistry Group at the University of Pennsylvania in 1965. The choice of topics of the papers and the order of presentation is intended to take the reader from the relatively easy, nonmathematical, introductory chapter, through four somewhat more difficult chapters dealing with various specific aspects of adsorption on solid electrodes, and a chapter devoted primarily to the role of solvent at the interface, to a final chapter presenting a highly sophisticated mathematical treatment of double layer theory.

Two common denominators may be found in the papers included in this book. One is the discussion of adsorption on solid electrodes as compared to adsorption on mercury, which has been much more extensively studied; and the other is the consideration of the process of electrosorption as distinct from gas phase adsorption and specifically the special role played by the solvent molecule in the former case.

The papers presented are not based primarily on original work of the authors, nor are they meant to be complete review articles. New ideas and new approaches to the presentation of the subjects may nevertheless be found in varying degrees in these papers. It is hoped that this volume will be useful to the ever-increasing number of scientists engaged in research in the expanding field of electrochemistry, the advanced student in electrochemistry, and the intelligent physical chemist who, hopefully, might gain some knowledge and become interested in the vast field of electrochemistry.

It is a pleasure to thank Professor Bockris for his encourgement during the preparation of this work and all my colleagues who spent long hours in preparation of the manuscripts. Thanks are also due Mr. C. Searles for drawing the diagrams.

Financial support during the preparation of the manuscript by the National Aeronautic and Space Administration, the National Science

Foundation, and the U.S. Army Engineer Research and Development Laboratory at Fort Belvoir is gratefully acknowledged. One of the authors (B.J.P.) wishes to thank the National Academy of Sciences, National Research Council, for the award of a postdoctoral research associateship at the Naval Research Laboratory at Washington, D.C.

Eliezer Gileadi

Contents

Chapter 3

Chapter 4

Chapter 5

The Potential of Zero Charge **87**

S. D. Argade and E. Gileadi

Chapter 6

The Role of Solvents at Electrodes. **117**

K. Müller

Chapter 1

Adsorption in Electrochemistry

E. Gileadi

1. INTRODUCTION

1.1. Special Aspects of Adsorption from Solution

In discussing the subject of electrosorption, i.e., adsorption upon electrodes in contact with electrolytic solution, one must first consider the special features of such processes, as opposed to adsorption from the gas phase.

The first and most important difference between adsorption from the gas phase and adsorption from solution is that the substrate surface is bare in the one case and solvated in the other. The process of adsorption from the gas phase can be represented by a simple bond formation with one or more sites (e.g., metal atoms) on the surface, with possible dissociation of the adsorbing molecule, e.g.,

$$H_{2_{gas}} \rightarrow H_{2_{ads}} \rightarrow 2H_{ads} \tag{1}$$

The intermediate $H_{2_{ads}}$ represents a physically adsorbed molecule. In many cases this intermediate state may not be stable and may pass, with essentially no energy of activation, to the final dissociatively chemisorbed state.

In comparison, the adsorption of hydrogen dissolved in water on the surface of an electrode would be represented as

$$H_{2_{soln}} + 2H_2O_{ads} \rightarrow 2H_{ads} + 2H_2O_{bulk} \tag{2}$$

In general, a physically adsorbed state replacing one or two water molecules could also occur, but in the case of hydrogen this would not be favored energetically.

1

Adsorption upon a solvated surface is, hence, a replacement reaction, and the effective standard free energy of adsorption ΔG^0_{ads} is given by the difference in free energies of adsorption of solute and solvent. Corresponding to equation (2) above, one has

$$\Delta G^0_{ads} = \left(2\mu^0_{H_{ads}} - \mu^0_{H_{2\,soln}}\right) - 2\left(\mu^0_{H_2O_{ads}} - \mu^0_{H_2O_{soln}}\right)$$

$$= \Delta G^0_{H_2} - 2\Delta G^0_{H_2O} \tag{3}$$

Thus, other things being equal, molecules will tend to adsorb on electrode surfaces in such a way that the least number of water molecules will be replaced. This tendency may, however, be offset by a stronger energy of bonding to the surface in one of the possible configurations.

The extent of adsorption also depends strongly on the solubility of the adsorbate. The lower the solubility, the higher the adsorbability. A systematic study [1] of the adsorption of several classes of organic compound on mercury (e.g., aliphatic alcohols, aromatic alcohols) revealed a linear relationship between the standard free energy of adsorption and the free energy of solvation within each homologous series. In a comparative study of adsorption of butanol and phenol on mercury from aqueous and methanolic solution [2], higher surface concentrations (at a fixed concentration in the bulk) were obtained when adsorption occurred from aqueous solution. Now, water is probably more strongly adsorbed on the surface of mercury than methanol, due to the smaller size of the molecule and the larger dipole moment. Thus, consideration of the *standard* free energy of adsorption would favor adsorption from methanol. The observed higher adsorption from aqueous solution in this case is, therefore, clearly due to the much lower solubility of the two alcohols tested in water. To eliminate the effect of different solubilities, the extent of adsorption should be compared at the same activity in solution, referred to the pure adsorbate (and therefore also to a saturated solution or a solution at equilibrium with a partial pressure of 1 atm,) as having unit activity.

A further important difference between adsorption on electrodes and gas phase adsorption is that in the former case the potential difference at the interface can be controlled and varied independently. In this way an additional degree of freedom exists and the equilibrium constant for adsorption can be changed at constant temperature and bulk concentration. Experimentally, this is evidenced by the now familiar "bell-shaped" dependence of coverage on potential [3] with the maximum in coverage occurring usually at potentials somewhat cathodic to the potential of zero charge. [4] Additionally, highly reproducible surfaces can be produced in rigorously purified solutions by keeping the electrode at a fixed potential [5] or taking it through a programmed cycle of potential variation [6].

2. EQUATIONS OF STATE AND ISOTHERMS

2.1. Definitions and Use

Of all the intensive variables of a system (e.g., pressure, temperature, density, molar volume), only a few are independent. For pure phases, fixing the values of two intensive variables determines unambiguously the state of the system. An equation relating any three intensive variables is called an equation of state. Usually the variables chosen are pressure, temperature, and molar volume or concentration, but other variables such as density, refractive index, dielectric constant, etc., could in principle be used.

The simplest equation of state for a bulk phase is the ideal gas law

$$PV = nRT \tag{4}$$

The corresponding equation of state for a two-dimensional ideal gas is

$$\Pi A = nRT \tag{5}$$

where Π is the surface pressure and A is the surface area. This is more usually written as

$$\Pi = \Gamma RT \tag{5a}$$

where $\Gamma = n/A$ is the number of moles per unit surface area. The surface pressure is related to the surface tension by the equation

$$\Pi = -(\gamma - \gamma_{\Gamma=0}) \tag{6}$$

or

$$d\Pi = -d\gamma \tag{6a}$$

where $\gamma_{\Gamma=0}$ is the surface tension at some reference state, chosen for convenience as that corresponding to zero surface concentration. The relationships in equations (6) and (6a) can be understood physically, since the surface pressure tends to increase the surface area while the surface tension tends to decrease it. Thus, the reversible work of increasing the surface area at constant Π is

$$W = \Pi \, dA \tag{7}$$

while the corresponding change in free energy is

$$\Delta G = \gamma \, dA \tag{8}$$

since by definition

$$\Delta G = -W \tag{9}$$

one has

$$\Pi \, dA = -\gamma \, dA \tag{10}$$

which is essentially equal to equation (6).*

The equation of state in two dimensions represents the relationship between the surface tension or excess surface free energy and the surface concentration. The adsorption isotherm, on the other hand, is the relationship between surface concentration and chemical or electrochemical potential, generally expressed as a function of the concentration in the bulk phase or the potential across the interface

$$\theta = Kf(c) \qquad \text{or} \qquad \theta = K'f(V) \tag{11}$$

The properties of the interface can be described equally well by an appropriate two-dimensional equation of state, or by the isotherm. The choice of way of representation is largely a matter of convenience, and depends usually on the type of data obtained experimentally. Thus, in the study of surface films on water, the surface pressure is the quantity measured directly, and the use of equations of state is a natural choice. The same applies to studies of adsorption on liquid metal electrodes, where surface tension data are obtained directly, and the surface concentration is calculable by use of purely thermodynamic arguments. In gas phase adsorption, and similarly in the study of adsorption on solid electrode, the surface concentration or the volume of gas adsorbed is directly obtained as a function of bulk concentration or pressure, and the adsorption behavior can be expressed more directly in the form of an isotherm.

---

* In writing equation (6) it has been arbitrarily assumed that $\Pi = 0$ at $\Gamma = 0$. The difference between equations (6) and (10) arises because in the former, Π represents the difference between Π at $\Gamma = \Gamma$ and Π at $\Gamma = 0$, while in the latter, the absolute value of Π at any Γ is used.

2.2. Conversion of Equation of State to Isotherms

The conversion of an equation of state, giving γ or π as a function of surface concentration Γ, to the corresponding isotherm where Γ is given as a function of bulk activity, is possible by making use of the Gibbs isotherm. The latter can be written at constant temperature and pressure as

$$d\gamma = -\sum \Gamma_i d\mu_i^\mu \tag{12}$$

The relationship between isotherms and equations of state can best be illustrated by specific examples [7]:

1. The equation of state for an ideal two-dimensional gas is

$$\Pi = \Gamma RT \tag{5a}$$

hence,

$$\frac{d\Pi}{d\Gamma} = -\frac{d\gamma}{d\Gamma} = RT \tag{13}$$

Equation (12) yields, for a binary system where the plane of reference at the interface is chosen such that $\Gamma_{solv} = 0$,

$$-\frac{d\gamma}{d\mu} = \Gamma \tag{14}$$

From equations (13) and (14) one obtains

$$\Gamma = K'a \tag{15}$$

or

$$\theta = Ka \tag{15a}$$

where a is the bulk activity of the solute and θ is the partial coverage $\theta = \Gamma/\Gamma_m$. Thus, the linear adsorption isotherm is equivalent to ideal gas behavior in two dimensions.

2. In a second example, the equation of state corresponding to the Langmuir adsorption isotherm will be derived.

Starting from the equation

$$\frac{\theta}{1-\theta} = Ka \tag{16}$$

one has

$$d \ln a = d \ln \left(\frac{\theta}{1-\theta}\right) = \frac{1}{\theta(1-\theta)} d\theta \tag{17}$$

Also

$$-\frac{d\gamma}{d\mu} = \Gamma \tag{14}$$

or

$$-\frac{d\gamma}{d\ln a} = RT\Gamma \tag{14a}$$

Combining equations (17) and (14a),

$$-d\gamma = \Gamma_m RT(1 - \theta)^{-1} d\theta \tag{18}$$

Integration of equation (18) and with the relationship given in equation (6a) yields

$$\Pi = -(\gamma - \gamma_{\theta=0}) = -\Gamma_m RT \ln(1 - \theta) \tag{19}$$

Equation (19) is the two-dimensional equation of state applicable to systems that obey the Langmuir isotherm.

A small inaccuracy has been involved in the above calculations. The quantity Γ used in the equations of state ($\Gamma = n/A$) and in the isotherms ($\Gamma = \Gamma_m\theta$) is the surface concentration of adsorbed material, while the same symbol in the Gibbs equation represents the surface excess. The error introduced by this approximation is, however, small in all cases of practical importance.

3. THE LANGMUIR ISOTHERM

3.1. Value and Deficiency

The Langmuir adsorption isotherm may now be regarded a classical law in physical chemistry. It has all the ingredients of a classical equation: It is based on a clear and simple model, can be derived easily from first principles, is very useful now, about fifty years after it was first derived [8] and will probably be useful for many years to come, and is rarely ever applicable to real systems, except as a first approximation.

3.2. Methods of Derivation

The simplest and most straightforward derivation is by the kinetic approach. Making the rate of adsorption proportional to the fraction of the surface which is unoccupied $(1 - \theta)$ and to the pressure or bulk concentration, and making the rate of desorption proportional to θ, one has, at equilibrium,

$$k_1(1 - \theta)p = k_{-1}\theta \tag{20}$$

hence

$$\frac{\theta}{1 - \theta} = Kp \tag{16a}$$

where

$$K = k_1/k_{-1} = \exp(-\Delta G^0_{\text{ads}}/RT) \tag{21}$$

Thermodynamically, the isotherm can be derived simply by applying the law of mass action to the adsorption process written as

$$M + X \rightleftharpoons MX \tag{22}$$

giving

$$\frac{a_{MX}}{a_M a_X} = K \tag{23}$$

where a_{MX}, a_M, and a_X are the activities of the adsorbed species on the surface, of the surface sites, and of the adsorbed species in solution, respectively. If all activities are now set proportional to the concentration in equation (23), equation (16) is obtained. Strictly speaking, the Langmuir isotherm *cannot* be derived thermodynamically unless some nonthermodynamic assumption is made regarding the values of the activity coefficients and their variation with the respective concentrations.*

In the statistical thermodynamic derivation, the $\theta/(1 - \theta)$ term arises from a probability term in the partition function related to the number of ways in which N^a indistinguishable particles can be arranged on N^s equivalent sites. Here again, one can only arrive at the Langmuir isotherm if it is assumed that the only θ dependent term in the partition function for the adsorbed species is the above probability term, i.e., that the rate of change of the total energy of the system with θ is constant.

The difference between localized and mobile layers is clearly made in the statistical derivation. The probability term in the first case will

---

* The same limitations apply if the isotherm is derived by equating the chemical potential μ_a of the adsorbed species to its chemical potential in the bulk. One can only write $\mu_a = \mu_a^0 + RT\ln(a_{MX}/a_M)$. The resulting isotherm depends on the function used to express a_{MX}/a_M. Writing $a_{MX}/a_M = \theta/(1 - \theta)$ essentially amounts to *assuming* the Langmuir isotherm rather than *deriving* it.

be replaced by the two-dimensional degree of freedom of translation in a total area A. This then leads to the simple linear isotherm

$$\theta = Ka \tag{15a}$$

It should be noted, however, that this result is only obtained if the area A available for the translational motion is assumed independent of θ. If the free area $A_F = A - sN^a = A(1 - \theta)$ is used in the partition function, where s is the area of one particle and N^a is the total number of particles on an area A, the Langmuir isotherm in its usual form [equation (16)] is obtained.*

3.3. Langmuir-like Adsorption

3.3.1. *Adsorption of Large Molecules.* The Langmuir isotherm as given above [equation (16)] was derived for adsorption of small particles, each occupying a single site on the surface. For dissociative chemisorption (e.g., H_2 in equilibrium with $2H_{ads}$), a slightly modified form of the same isotherm is applicable, namely,

$$\frac{\theta}{1 - \theta} = K^{1/2}p^{1/2} = k^1p^{1/2} \tag{16b}$$

For a molecule adsorbing on n sites on the surface without dissociation, one can write approximately

$$\frac{\theta}{(1 - \theta)^n} = Kp \tag{16c}$$

Now, the term $(1 - \theta)^n$ in equation (16c) arises from a consideration of the probability of finding n adjacent sites on the surface vacant. If the probability of any site being vacant is $(1 - \theta)$ and the sites on the surface are randomly occupied, equation (16b) results. However, it is clear that in the case of adsorption of large molecules, the sites on the surface are *not* randomly occupied, since it has already been assumed that they are occupied in groups of n sites having a particular fixed configuration, which depends on the shape of the adsorbate. Further-

---

* This is also consistent with the kinetic derivation. It could be argued that due to lateral free motion, all the surface is available for adsorption, independent of θ. However, at the moment of impact of a particle coming from the gas phase, the other particles on the surface may be considered stationary for a very short interval of time and the probability of the particle hitting a region on the surface not occupied at that instant is still $(1 - \theta)$.

more, the occupancy of adjacent sites is no longer independent under these conditions. The latter point will be illustrated for the simplest case of $n = 2$.

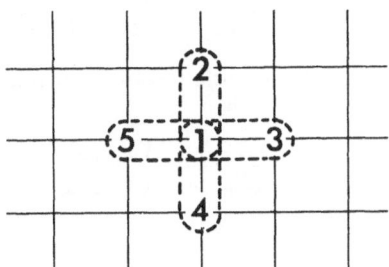

Considering a square array of sites as shown in the diagram, the site marked 1 can be occupied in one of four equivalent ways [sites (1,2) (1,3) (1,4), or (1,5)]. The probability of, e.g., site 2 being empty is $(1 - \theta)$. However, if it is known that site 2 is empty, one of the four equivalent ways of occupying site 1 is eliminated and the probability of it being also vacant is larger than $(1 - \theta)$. In other words, if the sites on the surface are occupied in groups of two, the vacant sites will also tend to occur in groups rather than randomly. The appropriate correction term to equation (16c) for the case of $n = 2$ was worked out by Miller [9] in 1939.

3.3.2. *Adsorption from a Second Layer.* A peculiar situation arises if physical adsorption on a second layer can occur along with chemisorption on part of the surface that is bare. The lifetime of a particle in the second layer depends on the energy of physical adsorption. If the energy of activation for lateral movement is very small (as would be expected for a physically adsorbed layer), the particle may have a relatively long average path on the surface before it evaporates; it may find a free site on the surface and become chemisorbed instead. This would lead experimentally to a linear adsorption isotherm up to relatively high values of θ [compared to that expected from equation (16) or (16b)] since particles hitting occupied sites on the surface will also end up being chemisorbed on a free site, and the rate of chemisorption will be nearly independent of θ [10].

3.4. Langmuir Adsorption from Solution

The adsorption behavior of organic molecules on electrode surfaces has recently been considered in detail by Bockris and

Swinkles [11] in terms of a replacement reaction [cf. equations (2) and (3)]. The equilibrium constant for the replacement reaction

$$R_{soln} + nH_2O_{ads} \rightleftharpoons R_{ads} + nH_2O_{soln} \tag{24}$$

can be written in terms of the mole fractions of reactants and products as

$$K = \frac{(X_{R,ads})(X_{W,soln})^n}{(X_{W,ads})^n (X_{R,soln})} \tag{25}$$

now in dilute solutions

$$X_{W,soln} \doteq 1 \qquad C_{R,soln} \doteq C_R/55.4 \tag{26}$$

and on the surface

$$X_{R,ads} = \frac{\Gamma_R}{\Gamma_R + \Gamma_W} \tag{27}$$

where Γ represents surface concentrations of the respective species.
Making use of the relationships

$$\Gamma_R = \theta\Gamma_{R,m} \quad \text{and} \quad \Gamma_{W,m} = n\Gamma_{R,m} \tag{28}$$

yields

$$X_{R,ads} = \frac{\theta\Gamma_{R,m}}{\theta\Gamma_{R,m} + (1-\theta)n\Gamma_{R,m}} = \frac{\theta}{\theta + (1-\theta)n} \tag{29}$$

and

$$X_{W,ads} = \frac{n(1-\theta)}{\theta + n(1-\theta)} \tag{30}$$

Substituting the values of the mole fractions from equations (26–30) into equation (25), one has

$$\frac{\theta}{(1-\theta)^n} \frac{[\theta + n(1-\theta)]^{n-1}}{n^n} = \frac{C_R}{55.4} K \tag{31}$$

Equation (31) should be compared with equation (16c) for adsorption from the gas phase. As expected, the equilibrium constant is modified by a quantity n^n dependent on the number of water molecules replaced. The coverage dependent term $[\theta + n(1-\theta)]^{n-1}$ is not very important and its variation with θ is quite small compared to the variation of the $(1-\theta)^n$ term, particularly at high coverage.

The result given in equation (31) is subject to the same limitations as equation (16c), which arise from the degree of order (or loss of randomness) due to occupation of sites on the surface in groups of n sites having fixed configurations. In the thermodynamic derivation of equation (31), the appropriate correction term should be introduced as an entropy factor affecting the activity of R_{ads}.

4. THE TEMKIN ISOTHERM

4.1. Assumptions, General Form, and Limiting Cases

The accumulation of experimental data showing a variation of the heat of adsorption with coverage has led Temkin to propose a new isotherm [12] based on the assumption of a linear variation of ΔH_{ads} and ΔG_{ads} with coverage:

$$\overline{\Delta G^0_{ads}} = \Delta G^0_{ads} + r\theta \tag{32}$$

where $\overline{\Delta G^0_{ads}}$ is the apparent standard free energy of adsorption and ΔG^0_{ads} is the standard free energy of adsorption of a bare surface $(\theta \to 0)$. The "Temkin parameter" r, which is often expressed in units of RT $(f = r/RT)$, is assumed here to be constant and independent of θ.

The Temkin isotherm is derived by assuming that the surface is made up of a great number of small patches ds, on each of which the Langmuir isotherm is applicable, but with the standard free energy of adsorption increasing in small steps. The isotherm is obtained by integrating the equation

$$d\theta = \frac{K(s)C}{1 + K(s)C} ds \tag{33}$$

where $K(s)$ is the equilibrium constant for adsorption (now a function of s) over the whole area, for s going from zero to unity. Expressing $K(s)$ as

$$K(s) = K_0 \exp(-rs/RT) \tag{34}$$

Temkin obtained

$$\theta = \frac{1}{f} \ln \left[\frac{1 + K_0 C}{1 + K_0 C \exp(-f)} \right] \tag{35}$$

The Temkin isotherm is best known in its approximate form

$$\theta = \frac{1}{f} \ln(K_0 C) \tag{35a}$$

applicable for intermediate values of the concentration and coverage and is usually referred to as the "logarithmic" isotherm. Equation (35a) is obtained from equation (35) for appreciable values of the parameter f when the concentration is low enough to satisfy the inequality

$$K_0 C \exp(-f) \ll 1 \tag{36}$$

yet high enough to have

$$K_0 C \gg 1 \tag{36a}$$

At very low concentrations

$$\theta = \frac{1}{f} \ln (1 + K_0 C) = K_0 C / f \tag{37}$$

and at very high concentrations $\theta \to 1$.

It was shown by Temkin [12] and others [13] that the same isotherm [equation (35)] can be obtained if the decrease in heat of adsorption is due to lateral interaction effects or to the dipole potential generated by the adsorbed species.

5. THE FRUMKIN ISOTHERM

5.1. Free Energy of Adsorption Decreasing with Coverage

The equation of state corresponding to the Langmuir isotherm was derived above:

$$\Pi = -RT\, \Gamma_m \ln (1 - \theta) \tag{19}$$

In 1925, Frumkin suggested [14] a modification of this equation to allow for long-range interactions between adsorbed species. Frumkin's equation of state has the form

$$\Pi = -RT\, \Gamma_m \left[\ln (1 - \theta) - \frac{f\theta^2}{2} \right] \tag{38}$$

The corresponding adsorption isotherm is

$$\left(\frac{\theta}{1 - \theta} \right) \exp(f\theta) = Ka \tag{39}$$

This is a valuable general isotherm from which both the Langmuir isotherm and the logarithmic Temkin isotherm can be obtained as special cases.

From equation (32) one obtains

$$K = K_0 \exp(-f\theta) \tag{34a}$$

Introducing this value of K into the Langmuir isotherm gives rise to equation (39). Thus, the Frumkin isotherm may be obtained directly by assuming that the apparent standard free energy of adsorption is a linear function of coverage. The variation of the partial coverage θ with the chemical potential of the species in the bulk phase (i.e., with log a) was calculated from the complete Frumkin isotherm [15] (equation 39) and is shown in Fig. 1 for various values of the parameter f. It is noted that while the term $\theta/(1 - \theta)$ cannot be considered negligible

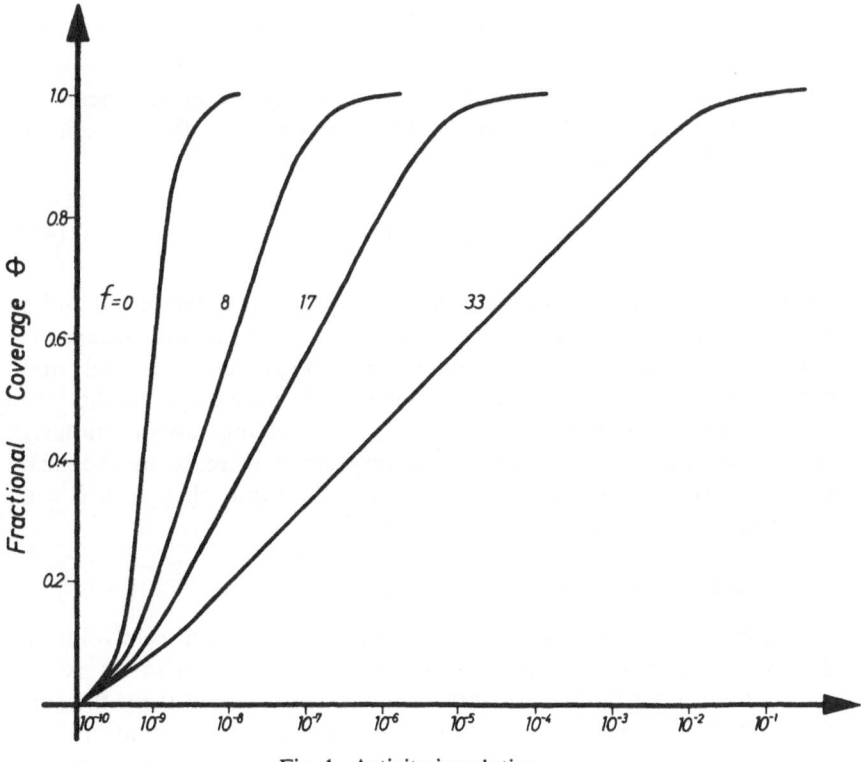

Fig. 1. Activity in solution.

compared to $\exp(f\theta)$ for any value of θ (unless an unreasonably high value of f is chosen), a region where θ is essentially a linear function of log a is obtained from $\theta = 0.1$–0.2 up to $\theta = 0.9$–0.8, depending upon the value of f. Further, the change in chemical potential corresponding to a change in θ from, e.g., $\theta = 0.2$ to $\theta = 0.8$ is characteristic of the numerical value of the parameter f and, in fact, can be used to measure this value experimentally.

5.2. Free Energy of Adsorption Increasing with Coverage

A case of interest occurs when the apparent standard free energy increases (in absolute value) with coverage [17]. Such behavior may be due to strong attraction forces between adsorbed species, which are particularly likely in the case of competitive adsorption.

Substituting $(-f)$ for f in equation (39), one has

$$\frac{\theta}{1-\theta} \exp(-f\theta) = Ka \tag{40}$$

Of the two terms on the left-hand side of equation (40), the first increases with θ while the second decreases. Differentiating the left-hand side of equation (40) and equating to zero, one obtains

$$\theta = 0.5 \pm 0.5 \left(1 - \frac{4}{f}\right)^{1/2} \tag{41}$$

Thus, for values of $f > 4$, the plot of θ vs. ln a has a maximum and a minimum at values of θ given by equation (41). The regions where the coverage decreases with increasing chemical potential are unstable and cannot be realized experimentally. Thus, as the coverage reaches the point where $\theta = 0.5 - 0.5(1 - 4/f)^{1/2}$, it will change discontinuously to a value close to unity due to a very small increase in chemical potential. When a is decreased gradually, the rapid change in θ will occur when $\theta = 0.5 + 0.5(1 - 4/f)^{1/2}$ [17].

5.3. Frumkin-Type Adsorption from Solution

It has been shown above that the heat of adsorption from solution differs from the heat of adsorption from the gas phase to an extent depending on the number of solvent molecules replaced from the surface and their energy of interaction with the surface [cf. equation (3)]. The solvent may also have a large effect on the entropy of adsorption.

Thus, the entropy usually decreases during adsorption from the gas phase, due to loss of two or three translational degrees of freedom. During adsorption from solution, several solvent molecules may be released from the surface for each molecule adsorbed and an increase in entropy may be expected. This has indeed been observed in several systems [3, 18].

The effect of the solvent on the rate of change of the apparent standard free energy of adsorption with coverage is complex and depends on the mechanism which causes the change in $\overline{\Delta G^0}$ with θ and on the type of bonding of the solvent molecules to the surface. Generally, the solvent will have a leveling effect and will tend to decrease the rate of change of $\overline{\Delta G^0}$ with θ. On a heterogeneous surface, the solvent molecules may adsorb irreversibly on the most active sites and render them unavailable for adsorption. Further, active sites may have a high energy of adsorption for solute as well as solvent molecules, while inactive sites will have a low energy of adsorption for both, so that the heat of adsorption will tend to be less dependent on the nature of the adsorption site.

6. EXPERIMENTAL TESTS OF THE ISOTHERMS

Up to this point, the adsorption isotherms were discussed in terms of the dependence of the partial coverage on the bulk activity a or the chemical potential μ. In electrosorption, the electrochemical potential $\bar{\mu}$ should be used as the independent variable. One may write

$$\bar{\mu}_A = \mu_A + F(V) \tag{42}$$

or

$$\bar{\mu}_A = \mu_A + G(q) \tag{42a}$$

where $F(V)$ and $G(q)$ are unspecified functions of the electrode potential or charge, respectively. The electrical variable affects the adsorption of charged species in a major way, but the adsorption of polar neutral molecules, and even those which do not have a permanent dipole moment, is also significantly dependent on potential or charge.

The adsorption of neutral organic molecules and of charged species is discussed in Chapters 6 and 7 of this book, respectively, in some detail. In the case of an adsorbed intermediate formed through charge transfer, equation (42) becomes [17]

$$\bar{\mu}_A = \mu_A{}^0 + RT \ln a_A + FV \tag{43}$$

and

$$\left(\frac{\partial \theta}{\partial \bar{\mu}}\right)_{T,P,a_A} = F \frac{\partial \theta}{\partial V} = \frac{F}{k'} C_{ps} \tag{44}$$

where C_{ps} is the adsorption pseudocapacity and k' is the charge associated with one complete adsorbed monolayer.

It was pointed out by Gileadi and Conway [17] that the dependence of θ on V was not very sensitive to the type of isotherm applicable to the system. The dependence of the adsorption pseudocapacitance C_{ps} on V was found to be very sensitive to the exact form of the adsorption isotherm (e.g., a small deviation from linearity in the dependence of $\overline{\Delta G^0}$ on θ can hardly be detected from the shape of the isotherm but is clearly seen as an asymmetry in the $C_{ps} - V$ plot), More recently, Parsons [19] considered several adsorption isotherms derived from equations of state corresponding to the hard sphere model in two dimensions with or without interaction, as well as the Langmuir and Frumkin isotherms. Parsons plotted θ vs. $\log(\beta a)$, where a is the activity of the adsorbent in solution and β is the adsorption coefficient defined as

$$\beta = \exp\left(-\frac{\overline{\Delta G^0}}{RT}\right) \tag{45}$$

where $\overline{\Delta G^0}$ is the apparent standard electrochemical free energy of adsorption [17], which may be a function of the charge or potential. It was found [19] that a plot of θ vs. $\log(\beta a)$ was not very sensitive to the type of isotherm employed, and a distinction between more complex two-parameter isotherms could only be made on the basis of comparison of plots of $d\theta/d \log(\beta a)$ vs. $\log(\beta a)$. Parsons also noted that the equations of state were even less sensitive than the adsorption isotherms to the nature of the forces involved in the adsorption process.

7. CONCLUSIONS

A great deal of knowledge and understanding of adsorption and heterogeneous catalysis has been gained by use of the Langmuir isotherm. This has sometimes been referred to as the "paradox of heterogeneous catalysis" [20], since conditions in many cases were such that significant deviations from Langmuir behavior occurred. The state of the field of heterogeneous kinetics and, in particular, electrode kinetics has reached a point where a refinement of theory is called for. This can be achieved by considering a linear variation of the heat or the apparent standard free energy of adsorption with coverage, giving rise to the Frumkin isotherm. If the Langmuir isotherm may be considered a zero-order approximation to real situations in cases where the coverage is intermediate, the Frumkin isotherm is a first-order approximation. Thus, systems may be considered which exhibit Langmuir behavior with small deviation (i.e., $\overline{\Delta G}^0$ changes only slightly with θ) or Frumkin-type behavior, where $\overline{\Delta G}^0$ changes a lot with θ but in an approximately linear manner. Detailed complex isotherms have been applied to describe adsorption on mercury [1, 21-23]. These, however, are only useful for liquid metals where the charge on the metal can be determined directly and the coverage can be measured with a relatively high degree of accuracy. At the present stage of knowledge of adsorption on solid electrodes, where even the real surface area, and hence the partial surface coverage, are only known with 20 to 50 % accuracy, further refinements over the Frumkin isotherm do not seem necessary.

For solid electrodes, consideration of the isotherms is preferable to discussion in terms of equations of state, since the form of the isotherm can be associated more directly with the atomic mechanism of adsorption and also since the isotherm is experimentally determined. For liquid electrodes, the surface tension and its variation with potential can be measured directly and accurately. Here, equations of state may be used but, as pointed out by Parsons [19], they are less indicative of the type of adsorption taking place and a distinction between equations of state based on different physical models may not be possible.

The effect of solvent in adsorption on electrodes is very important. It modifies the heat and entropy of adsorption in that adsorption becomes a substitution reaction. It also diminishes the apparent rate of change of heat of adsorption with coverage and brings about a potential dependence of adsorption due to the variation of the standard free energy of adsorption of the solvent with potential and due to the difference in dielectric constant between solvent and solute.

REFERENCES

1. Blomgren, Bockris, and Jesch, *J. Phys. Chem.* **65**, 2000 (1961).
2. Bockris, Muller, and Gileadi, *Electrochim. Acta*, in press.
3. Gileadi, Rubin, and Bockris, *J. Phys. Chem.* **69**, 3335 (1965).
4. Bockris, Green, and Swinkels, *J. Electrochem. Soc.* **111**, 743 (1964).
5. Reddy, Genshaw, and Bockris, *J. Electroanal. Chem.* **8**, 407 (1964).

6. Gilman, *J. Phys. Chem.* **67**, 78 (1963); *Electrochim. Acta* **9**, 1025 (1964).
7. Swinkels, Ph. D. Dissertation, University of Pennsylvania (1963).
8. Langmuir, *J. Am. Chem. Soc.* **40**, 1361 (1918).
9. Miller, *Proc. Cambridge Phil. Soc.* **35**, 293 (1939).
10. Hayward and Trapnell, *Chemisorption*, 2nd ed., Butterworths, Washington, D. C. (1964).
11. Bockris and Swinkels, *J. Electrochem. Soc.* **111**, 736 (1964).
12. Temkin, *Zh. Fiz. Khim.* **15**, 296 (1941).
13. Horiuti, *Trans. Symposium on Electrode Processes*, Yeager, ed., John Wiley & Sons, New York, p. 17 (1959).
14. Frumkin, *Z. Physik. Chem.* **116**, 466 (1925).
15. Conway and Gileadi, *Trans. Faraday Soc.* **58**, 2493 (1962).
16. Conway, Gileadi, and Dzieciuch, *Electrochim. Acta* **8**, 143 (1963).
17. Gileadi and Conway, *Modern Aspects of Electrochemistry*, Vol. III, Chap. 5, Bockris and Conway, eds. Butterworth, Washington (1964).
18. Heiland, Gileadi, and Bockris, *J. Phys. Chem.* **70**, 1207 (1966).
19. Parsons, *J. Electroanal. Chem.* **7**, 136 (1964).
20. Boudart, *J. Am. Inst. Chem. Engr.* **2**, 62 (1956).
21. Conway and Barradas, *Electrochim. Acta* **5**, 319 (1961).
22. Bockris, Devanathan, and Muller, *Proc. Roy. Soc. (London)* **A274**, 55 (1963).
23. Wroblowa, Kovac, and Bockris, *Trans. Faraday Soc.* **61**, 1523 (1965).

Chapter 2

Organic Adsorption at Electrodes

B. J. Piersma*

1. INTRODUCTION

The types of information required for a discussion of adsorption of gases at metal surfaces have been enumerated by Bond [¹]. With minor variations, similar information is desired from the measurements of adsorption of organic compounds on solid metal electrodes. This includes:

1. The conditions of adsorption; i.e., under what conditions of temperature, pressure, or concentration, etc. and on which metals does the compound in question chemisorb?

2. The stoichiometry of the adsorption process; is chemisorption associative or dissociative, how many sites are involved for the adsorption of one molecule, is the adsorbed species mobile on the surface, what is the nature of bonding?

3. The energetics of adsorption; the equilibrium constant and its temperature variation, free energy, heat and entropy of adsorption, their variation with coverage, the activation energy for adsorption and desorption, the strengths of the bonds involved in the adsorption process.

4. The adsorption isotherm; the variation of coverage with pressure or concentration, temperature, potential, solvent, and pH.

5. The kinetics of adsorption; the rates of adsorption and desorption. With the only recently developed interest in adsorption at solid metal electrodes, it should be obvious that very little of this information has been obtained for even one system. As a matter of fact, the desired

* Present address: Eastern Baptist College, St. Davis, Pennsylvania.

information is far from complete for even a single system in gas phase adsorption [1, 2]. Therefore, any discussion of organic adsorption at solid electrodes must necessarily be incomplete with respect to experimental evidence and information. In this chapter we shall present much of the available data on the electrosorption of organic compounds, although no attempt is made at completeness. Information from adsorption on mercury, recently reviewed by Frumkin and Damaskin [3] and gas phase adsorption are presented where comparison seems justified.

2. COMPARISON OF METHODS FOR ADSORPTION STUDIES IN THE GAS PHASE AND AT ELECTRODES IN SOLUTION

The experimental methods for the study of electrosorption have been reviewed for equilibrium adsorption [4] and for adsorbed intermediates [5]. Methods applicable in gas phase studies have been reviewed by Bond [1] and by Trapnell [2], and modern techniques are discussed by Ehrlich [6]. The fundamental difference of the solution phase presents severe limitations on the experimental approach to electrosorption, and many of the exciting techniques now being utilized in the gas phase are simply not applicable. Table I lists the various methods that can be used to study adsorption in the gas phase and at electrodes in solution. In gas phase study, tremendous advances have

Table I. Methods for the Study of Adsorption

Gas phase	Solid electrodes in solution
1. Volummetry and gravimetry	1. Coulometric: galvanostatic charging, galvanostatic hydrogen deposition, potential sweep, chronopotentiometry
2. Accomodation coefficients	
3. Magnetic susceptibility	
4. Specific magnetization	
5. Work function changes	2. Differential double layer capacity
6. Electrical conductivity	3. Radiotracer: isotope-labeling detection in solution or on the electrode
7. Kinetic studies—isotopic traces	
8. Infrared spectroscopy	4. Spectrophotometric: change in solution absorption of ultraviolet
9. Gas–solid chromatography	
10. Electron spin resonance	5. Ellipsometry
11. Field ion microscopy	
12. Field emission microscopy	
13. Electron diffraction, LEED	
14. Flash desorption	

been made, due primarily to the development of ultrahigh vacuum techniques. For the first time it has become possible to specify exactly the environment at an interface and to perform meaningful measurements on samples with small surface areas [6]. The use of high vacuum has permitted the use of the field emission microscope, which can scan the surface on a scale that approaches the realm of atomic dimensions, and the field ion microscope, which gives an evenly increased resolution. In many cases new problems arise in the study of clean systems. The structural studies of clean surfaces with low-energy electron diffraction (LEED) have caused more problems than they have solved [7], and at present adsorption is being investigated before the clean surfaces themselves are understood.

For the study of adsorption, two requirements should be met: Both the metal surface and adsorbing species must be in known reproducible states and the environment must be adjusted to prevent introduction of impurity. The speed of contamination of clean metal surfaces has been realized for some time in gas phase studies [8]. In the gas phase, the time for 1 % contamination of the metal surface, assuming one out of every four molecules striking the surface will stick, is given by [6]

$$\tau = (10^{-7}/p) \text{ sec}$$

Thus, at 10^{-6} torr, 100 msec are available for experimentation with a metal surface less than 1 % contaminated. The above requirements have been fairly well met in a few studies by the use of ultrahigh vacuum techniques, and the problem is now well understood for gas phase adsorption measurements.

The study of adsorption in solution presents somewhat more of a problem. Although the influence of impurity was realized by Frumkin's school in the 1930's and reemphasized by Bockris and co-workers, acceptable conditions of purity have not in general been maintained in electrochemical systems. A recent and rather unfortunate trend in electrochemical measurements has been to substitute transient methods and electrode pretreatment sequences for the design and maintainance of clean systems. Just as gas phase studies required the development of the Bayard–Alpert inverted ionization gauge to measure pressures in ultrahigh–vacuum systems, perhaps methods for testing the purity of solution and electrode cleanliness are required before consistent and meaningful measurements at electrodes in solution are made. A step in this direction has been made by Schuldiner and Warner [9], who have suggested a simple method for determining the cleanliness of a platinum anode. Using the equivalent of high–vacuum techniques (electro-chemical measurements in solution can never approach ultrahigh-vacuum techniques for obvious reasons), it has been possible to main-tain highly clean systems for periods of several months [10]. These results show that with proper system design and a little effort, electro-chemical measurements can be made in clean systems reproducibly without resorting to techniqes that act as substitutes for system purity. Thus the design and operation of clean electrochemical systems would seem to be a prerequisite before reproducible and indeed meaningful information on electrosorption of organic compounds can be obtained. The often-heard argument that the organic species being examined is itself an impurity and therefore system cleanliness is impossible is not relevant.

The question of which method is best for the study of electro-sorption may be simply answered by stating that all of the available methods have very definite limitations, and in the absence of a really good method it is wise to employ all of the available techniques that give complementary information. The experimental problems involved in radio-tracer techniques are relatively greater; however, the results are more easily interpreted in terms of total amount of species formed and based on fewer assumptions than are the various electrochemical methods. Other methods have been attempted, e.g., infrared spectro-scopy using internal reflection and electron spin resonance [11]—ho-wever, with little success. Electrosorption is still awaiting the develop-ment of a good *in situ* method of examining adsorbed layers at metal surfaces.

3. ENERGETICS OF ADSORPTION

The intention for this section is to compare the thermodynamic quantities determined for the adsorption of selected compounds in the gas phase with those for the electrosorption of the same species. This comparison is at best difficult since gas phase studies have in general determined the heat of adsorption primarily from calorimetric measure-ments or from isotherms using the Clausius–Clapeyron relation, while the few studies of electrosorption which have considered the energetics have determined free energies from isotherms. Heats of adsorption have been determined at platinum electrodes for ethylene [12] and benzene [13] from the variation of equilibrium constant, determined from the adsorption isotherms, with temperature using the van 't Hoff equation. However, the adsorption of these compounds on platinum in the gas phase has not been reported. Hydrogen adsorption, which has been extensively studied in the gas phase [2] on a number of metals, has been examined at Pt, Rh, and Ir electrodes by Breiter [14] using the technique of cyclic voltammetry to obtain free energy, heat, and entropy of adsorption. While this is not organic adsorption, it can give an indication of the differences in gas phase processes and electrosorption. Heats in the gas phase range from -45 kcal/mole on tungsten to -26 kcal/mole on palladium and -28 kcal/mole on rhodium [2]. On platinum and rhodium, Breiter has obtained heats of adsorption of about -18.5 kcal/mole at zero coverage and from -8.5 to -13.0 kcal/mole at $\theta = 0.5$. These values tend to show that the heats of adsorption on electrodes in solution are in general much less than those in the gas phase, which is to be expected for the replacement reaction [15]. Limitations on the quantities determined by Breiter have been discussed by Frumkin [16] so that these values should be used with caution. In addition, Breiter has assumed the Nernst relation to determine the partial pressures of hydrogen from electrode potentials. This assumption has been shown not to be valid below 10^{-6} atm hydrogen partial pressure [10] because of hydrogen absorbed in the Pt and becomes invalid at somewhat higher partial pressures in the presence of oxygen [17].

Table II shows the difficulty in making a comparison of gas phase and electrosorption measurements. In general, the heats of adsorption

Table II. Heats of Adsorption, kcal/mole)

Element	Gas phase			Electrosorption		
	$H_2^{(2)}$	$CO^{(1,2)}$	$C_2H_4^{(1,2)}$	$H_2^{(14)}$	$C_2H_4^{(12)}$	$C_6H_6^{(13)}$
W	-45	-82	-102			
Ni	-30	-42	-58			
Fe	-32	-46	-68			
Rh	-28		-50	-18.5		
Pd	-26					
Pt				-18.5	0	$+10$
Cu		-9	-19			
Au		-9	-21			

are much lower in electrosorption. This has been explained by the fact
that electrodes are covered with water molecules and energy is required
to displace them from the surface. The idea that electrosorption is a
replacement reaction is more clearly brought out by examining the free
energy quantities determined for the adsorption of a number of organic
species at electrodes. The net free energies of adsorption—defined as [18]

$$\Delta\bar{G}_a{}^0 = (\bar{\mu}_A^{0,a} - \bar{\mu}_A^{0,s}) - n(\bar{\mu}_{H_2O}^{0,a} - \bar{\mu}_{H_2O}^{0,s})$$

where $\bar{\mu}_A^{0,a}$ and $\bar{\mu}_{H_2O}^{0,a}$ are the standard electrochemical potentials of
adsorbate and water in the adsorbed state, referred to unit mole
fractions on the metal surface, $\bar{\mu}_A^{0,s}$ and $\bar{\mu}_{H_2O}^{0,s}$ are the standard electrochemical potentials of adsorbate and water in solution, referred to unit
mole fractions in the solution, and n is the number of water molecules
replaced, determined on mercury by the electrocapillary method—are
given for a number of butyl, phenyl, and naphthyl derivatives in
Table III. The standard electrochemical free energy of adsorption as
obtained from the Langmuir isotherm in the form

$$\frac{\theta}{1-\theta} = \frac{C_A}{55.5} e^{-\Delta\bar{G}^0/RT}$$

has been determined for the adsorption of several organic bases and
their conjugate acids on mercury [19], and these are presented in Table IV.
Finally, the standard free energies of adsorption for ethylene and
benzene determined from equilibrium constants [12, 13] and for

Table III. Net Free Energies of Adsorption on Hg [18]

(0.1 N HCl, potential at ecm,* $\theta = 0.25$)

Functional group	$-\Delta\bar{G}_a{}^0$, kcal/mole		
	$N-C_4H_9$	C_6H_5	$C_{10}H_7$
OH	3.7	5.6	8.8
CHO	6.5	6.6	9.0
COOH	4.5	6.7	8.6
CN	4.5	6.0	
$NH_3{}^+$	2.9	3.9	7.3
SH	4.9	7.5	
$SO_3{}^-$	2.6	3.8	7−7.5
CO	~8		

* Electrocapillary maximum.

Table IV. Comparison of Free Energies of Adsorption of
Organic Bases [19]
(potential -600 mV vs. SCE,* 26°C)

Adsorbate	$-\Delta\bar{G}^0$, kcal/mole 1 N HCl $\theta = 0.5$	$\theta = 0.25$	$-\Delta\bar{G}^0$, kcal/mole 1 N KCl $\theta = 0.5$	$\theta = 0.25$
Pyridine	2.5	4.8	3.3	
2-NH$_2$-pyridine	4.5	5.3	4.0	
2-Cl-pyridine	3.5	5.6	5.3	5.9
1,2,3,6-tetrahydro-pyridine	3.7	4.6	4.5	
Piperidine	3.6	4.3	4.6	
Aniline	4.1		4.5	

* Saturated calomel electrode.

n-decylamine and naphthalene determined from the following iso-therm [20, 21]

$$\frac{\theta}{(1-\theta)^n}\frac{[\theta + n(1-\theta)]^{n-1}}{n^n} = \frac{C_{org}}{55.5}e^{-\Delta\bar{G}_a^0/RT}$$

which reduces to the usual Langmuir isotherm when n, the number of water molecules displaced by one organic molecule on adsorption, becomes equal to 1, are given in Table V for solid metal electrodes. It is

Table V. Standard Free Energies of Adsorption at Potential of
Maximum Adsorption, $\theta - 0$

Metal	$-\Delta\bar{G}^0$, kcal/mole				
	A	B	C	D	E
Ni	6.0	6.8	—	—	—
Fe	7.0	6.6	—	—	—
Cu	7.0	7.3	—	—	—
Pd	—	6.2	—	—	—
Pt	8.4	7.4	5.6	5.7–8.7 [13]	—
Au	0.7*	—	—	6.1 [22]	<5.8

A: naphthalene, 1 N NaClO$_4$ [21].
B: n-decylamine, 0.9 N NaClO$_4$, 0.1 N NaOH [20].
C: ethylene, 1 N H$_2$SO$_4$ [12].
D: benzene, 1 N H$_2$SO$_4$.
E: cyclohexane, 1 N H$_2$SO$_4$ [22].
* 1 N H$_2$SO$_4$ [22].

readily apparent that the standard free energies of adsorption of a wide variety of organic compounds on a number of solid metal electrodes and on mercury are quite similar with a variation of less than 10 kcal/mole and in most cases less than 5 kcal/mole. This leveling effect in electrosorption is thus a good indication of the strong influence of solvent molecules in determining adsorption energies. There are also some indications that the presence of water vapor in gas phase adsorption can have large effects on the heats of adsorption [1, 2]. The effect of the adsorption of water vapor on contact potentials has been known for some time, and particularly large effects are obtained in critical ranges of water vapor below 1 % and above 75 % [23]. Thus, electrosorption is most likely similar to competitive gas phase adsorption when water vapor is present and in this sense is not a special type of adsorption.

Thermodynamic quantities other than the standard free energy of electrosorption have been worked out only for the cases of ethylene [12] and benzene [13] adsorption on Pt electrodes. Using a thermodynamic cycle of the type

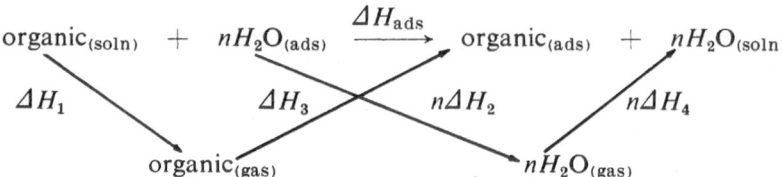

and appropriate values for the various enthalpy quantities, the calculated heats of adsorption for the overall process were in good agreement with the observed values. The entropies of adsorption were calculated in the usual way from the free energy and enthalpy values. These values are collected in Table VI. It is noted that the entropies of adsorption are positive, as a result of the desorption of water molecules.

Further comparison of the energetics of adsorption from the gas

Table VI. Thermodynamic Quantities for Adsorption of Ethylene and Benzene on Pt Electrodes

	$\Delta \bar{G}_a^0$, kcal/mole	ΔH, kcal/mole	ΔS, eu
C_2H_4	5.6	~0	+16
C_6H_6	5.7–8.7	+10	+50

phase and of electrosorption must await more experimental data from both types of systems. In general, comparisons will have to be restricted to specific systems. For example Smith and Burwell [24] have suggested that results for liquid phase hydrogenation closely resemble those in the vapor phase, and it appears unlikely that there is any substantial difference between the mechanism of hydrogenation on Pt catalysts in the vapor phase and that in the liquid phase. Bond and Wells [25] have stated that on energetic and steric grounds it should be expected that certain processes will occur on a nearly bare surface while the same processes would be unable to proceed on a highly covered surface.

4. TRENDS IN ELECTROSORPTION

From the large number of adsorption measurements made at Hg electrodes, definite trends for the electrosorption of organic molecules in general can be suggested. These are taken primarily from the summary of Frumkin and Damaskin [3] and the papers of Barradas and Conway [19] and of Bockris and co-workers [18]. Further trends, when evident, from adsorption measurements at solid electrodes will also be given [12, 13, 20–22].

Electrocapillary measurements show that the decrease in interfacial tension in organic systems reaches its maximum in the vicinity of the potential of zero charge (pzc). Of course, this is related to the "bell-shaped" curve for the variation of coverage with potential observed for adsorption at solid electrodes, particularly at lower solute concentrations. This has been explained in terms of the water competition model of Bockris, Devanathan, and Muller [26]. More recently, Devanathan has objected to this model [27] and attributed the displacement of organic species by water molecules to the hyperpolarizability of water. These authors also suggest that any substance with a dielectric constant greater than that of water in the double layer will displace water even at high field strengths. Frumkin has recently proposed [28] that the potential dependence of organic adsorption on Pt is related to competition with hydrogen at low potentials and oxygen at high potentials and not primarily with water molecules. This theory would require oxygen species to be present on the electrode beginning at about 0.6 V_{NHE} (NHE = Normal hydrogen electrode) in sufficient amounts to compete with organic species.

While the exact reason for the potential dependence of organic adsorption can remain open for discussion, the fact of this dependence does remain and is typical of electrosorption of neutral organic molecules. The adsorption maximum is in general slightly negative to the pzc, which can be interpreted, at least by the water competition model, as due to the slightly preferred orientation of water molecules with oxygen toward the metal surface at the pzc. From some measurements on Hg where the pzc is well established, there are good indications that the adsorption maximum is on the negative side for aliphatic molecules but on the positive side for aromatic molecules [18]. The pzc was displaced to positive values by the adsorption of aliphatic compounds and to negative values by aromatics. This shift in pzc is due to the substitution of organic molecules for water dipoles. From adsorption studies of aromatic and cyclic bases, it was reported [19] that no sharp maximum in adsorption near the pzc is observed as compared with aliphatic species. The suggestion that this was not related to the π-orbital character of the rings was made since hydrogenated derivatives of pyridine maintained the tendency.

The shift of the pzc, in a negative direction and the adsorption of aromatic compounds on positively charged Hg surfaces are attributed to Π-electron interaction, which is facilitated by the flat orientation of the benzene ring. Neutral organic molecules that do not contain π-electrons were not found to adsorb on Hg at large positive surface charges. Adsorption on positive surfaces was found to increase as the number of double bonds or π-electrons increased. A comparison of electrocapillary curves measured in solutions of organic substances forming an homologous series showed that the decrease in interfacial tension and consequently adsorption at identical concentrations increases with an increase in chain length. Adsorption increases with increasing concentration of solute and also increases with concentration of the supporting electrolyte. This latter effect has been attributed to a salting out effect and not to adsorption of inorganic ions. Adsorption is strongly reduced by competitive adsorption of anions. Surface-active ions can either increase or decrease organic adsorption. The adsorption of anions decreases adsorption of neutral molecules but can increase the adsorption of organic cations by decreasing the mutual repulsive forces on the surface. From an examination of the simultaneous adsorption of n-butyl alcohol and tetralkylammonium cations, it was found that desorption of the alcohol was less pronounced with the larger cations, being in the sequence

$$K^+ > (CH_3)_4N^+ > (C_2H_5)_4N^+ > (C_4H_9)_4N^+.$$

From electrocapillary measurements it can be concluded that in general aliphatic organic molecules are oriented perpendicular to the metal surface while aromatic molecules are oriented parallel, i.e., adsorbed flat, on the metal surface. In neutral solutions, the values of limiting adsorption for aniline, pyridine, and their derivatives correspond to an orientation of adsorbed dipoles perpendicular to the surface. Butyl derivatives were found to adsorb perpendicular with the hydrocarbon end of the molecule toward the metal [18]; however, it has been reported that for aliphatic oxygen compounds the molecule is oriented with the hydrocarbon end toward the interface and the negative end toward the metal [3]. Obviously the orientation of the dipoles will generally depend on the charge on the metal surface. It has also been proposed that the orientation will depend on the charge; e.g., the vertical orientation of molecules, which is characteristic of negatively charged surfaces can be substituted by a horizontal one, in which case the π-electrons of an aromatic ring can interact with the positive charges of the Hg surface. In terms of energy changes, the standard free energy of adsorption was found to change symmetrically about the pzc for some organic species, e.g., ethylene, but to change unsymmetrically for others, e.g., aromatic bases, where $\Delta \bar{G}^0_{ads}$ was lower at more anodic potentials. In general, the free energy increases with coverage, i.e., becomes more negative, on negatively charged surfaces, and decreases with increasing coverage on positively charged electrode surfaces. Conway [19] was able to obtain a linear correlation of $\Delta \bar{G}^0_{ads}$ with $\theta^{3/2}$ but with sharp inflections, which were attributed to reorientation of dipoles as θ increased and to condensing of the dipole film.

A comparison of differential capacity curves in the presence of organic species shows a similarity in the first approximation of those obtained on solid metals to those obtained with Hg. A decrease in double layer capacity with adsorption is observed, and the organic species is seen to desorb with increasing electrode charge. The peaks on the capacity–potential curves were less pronounced on solid metals than on Hg. With an increase in chain length of the adsorbing organic molecule, the double layer capacitance is observed to decrease due to an increase in thickness of the double layer.

In terms of isotherms, it has been suggested that the Temkin isotherm is only of limited applicability to the adsorption of neutral organic molecules [3]. This statement is supported by the fact that the changes in energies and heats of electrosorption are relatively small, being on the order of a few kilocalories per mole. The point concerning the applicability of the Temkin isotherm to organic adsorption at electrodes has been further treated in this light [29].

A number of trends are also readily available from adsorption studies in the gas phase. For example, alkynes and dienes are generally very strongly adsorbed on all metals, and it is assumed that these are π-bonded to the surface. Metals which form ethylene complexes will also form complexes with other olefins and vice versa. The extent of olefin isomerization and exchange is always characteristic of the metal and substantially independent of molecular weight. In addition, correlations have been made between the organometallic chemistry of transition metals and their adsorption properties and catalytic chemistry [25]. For example, π-olefin complexes are formed by virtually all group-VIII metals, and the strength of olefin-metal bonds may vary considerably, depending on other ligands present in the complex. All these metals adsorb olefins and are active in olefin hydrogenation. Both in catalytic and organometallic chemistry, an acetylene molecule can displace an olefin molecule unless the latter is either adsorbed at a metal surface or functioning as a ligand in a complex. In general, such correlations are not as easily applicable to electrosorption as, for example, are the trends in adsorption on Hg, and so these will not be further discussed. It would be very instructive to compare such trends in gas phase adsorption obtained in the presence of water vapor, i.e., under conditions of competitive adsorption that apply in electrosorption, but unfortunately such information is not available.

5. MODES OF ADSORPTION AND STRUCTURE OF ADSORBED SPECIES

Although a relatively large number of studies recently have been concerned with organic adsorption at solid electrodes, the information obtained in general has been rather meagre. Table VII gives a fairly complete listing of systems that have been studied and the experimental methods used. In general, the aim has been to obtain coverages using coulombic methods, which are subject to several limitations. The information on coverages has been discussed elsewhere [4] and will be reviewed only briefly here for completeness. The aim in this discussion will be to examine information concerning the adsorbed species and its comparison with relevant gas phase studies. This can best be accomplished by dividing the information into two sections: that resulting

Table VII. Information on Electrosorption of Organic Compounds on Solid Metal Electrodes

System	Electrolyte	Temperature, °C	Information obtained	Method*	Reference
CO, Pt	1 M H$_2$SO$_4$	25	Q at open circuit	A	30
CO, Plat. Pt	1 N H$_2$SO$_4$	15–50	Q as $f(V, T)$	A	31
CO, Pt	1 N HClO$_4$	20	Γ as $f(V)$	B	32
CO, Pt	1 N HClO$_4$	30	θ as $f(V)$	B	33
CO, Pt	1 N HClO$_4$	40	Q, θ_H or $f(V)$	A, C	34
HCOOH, Pt	1 N HClO$_4$	30	θ as $f(V,$ conc)	A, C	35
HCOOH, Pt	1 M HClO$_4$+NaOH	40	θ as $f(V)$	A, C	36
HCOOH, Plat. Pt	5 N H$_2$SO$_4$	25–90	θ as $f(V)$	C	37
HCOOH, Pt	1 N HClO$_4$	40	θ as $f(V,$ conc)	A	38
HCOOH, Plat. Pt	1 M H$_2$SO$_4$	25	Q	B	39
HCOOH, Pt	1 N HClO$_4$	25	θ as f(time, conc)	B	40
CH$_3$OH, Pt	1 N NaOH	25	θ as f(conc)	A	41
CH$_3$OH, Pt	1 N HClO$_4$	25	θ as $f(V,$ conc)	B, C	42
CH$_3$OH, Pt	1 N HClO$_4$	25	rate of ads	B	43
CH$_3$OH, Pt, Ir, Au, Pd, Rh	1 N HClO$_4$	30	Capacity as f(conc)	B	44
CH$_3$OH, Plat. Pt	0.1 N H$_2$SO$_4$		Q as $f(V$ open circuit)	A	45
CH$_3$OH, Plat. Pt	H$_2$SO$_4$, KOH		Q as f(pH)	A	46
CH$_3$OH, Plat. Pt	0.1 N H$_2$SO$_4$	20	θ as $f(V)$	B	47
CH$_3$OH, Plat. Pt	1 N H$_2$SO$_4$		Q	A	48
CH$_3$OH, Pt	1 N H$_2$SO$_4$	25	θ as $f(V,$ conc)	D	49
CH$_3$OH, Plat. Pt	H$_2$SO$_4$, KOH	20	Q as f(pH)	A, B	50

* A: anodic galvanostatic charging.
  B: linear anodic potential sweep.
  C: cathodic galvanostatic charging.
  D: linear cathodic potential sweep.
  E: a-c impedance.
  F: differential capacitance from galvanostatic pulse.
  G: ellipsometry.
  H: ultraviolet absorption in solution.
  I: radiotracer.
  J: volumetric.
  K: potential decay.

Table VII *(continued)*

System	Electrolyte	Temperature, °C	Information obtained	Method*	Reference
Pt-Ru, Pd-Ru, CH_3OH, Plat. Pt	H_2SO_4, KOH	20	Q as f(pH)	A	51
CH_3OH, Pt	1 N H_2SO_4	25	Q	A, B	52
CH_3OH, Pt-Ru	H_2SO_4, KOH	20	Q as f(pH)	A	53
C_2H_5OH, Plat. Pt	1 N H_2SO_4		capacity as $f(V, conc)$	E	54
C_2H_5OH, Plat. Pt	0.1 N H_2SO_4	20	Q as $f(V, conc)$	A, B	55
C_1–C_5 alcohols, Au	1 N $HClO_4$	5–25	capacity as $f(V)$	F	56
Allyl alcohol, allyl methanol, acrolein, methacrylic acid	H_2SO_4, KOH, Pd		Q	C	57
CH_2CHCH_2OH, Plat. Pt	0.1 N H_2SO_4		Q, capacity as f(conc)	A, C	58
Butane-1-4-diol and derivatives	Pt		Q, Γ	A	59
$(COOH)_2$ Au	H_2SO_4		film thickness	G	60
Pyridine, acridine quinoline, polyvinylpyridine	Aq and methanol Cu, Ag, Ni	25	or f(conc)	H	61
Hydroquinone phenylenediamine methylaminophenol	Na_2CO_3Ag	20, 40	θ as f(conc)	H	62
Thiourea Au	$H_2SO_4 - Na_2SO_4$	25	θ or f(conc) energetics	I	63
Amyl alcohol caprylic acid diphenylamine	Pt 1 M $HClO_4$		capacity as $f(V, conc)$ θ as $f(V, conc)$	B, E	64
C_2H_4, Pt	1 N NaOH	60	Γ as $f(V, time)$	I	65
C_2H_4, Plat. Pt	1 N H_2SO_4	25	Q	B	66
C_2H_4, C_2H_6, Plat. Pt	1 N H_2SO_4 1 N KOH	21–97	Q, θ as $f(T)$	A	67
C_2H_4, Pt	1 N H_2SO_4	30–70	θ as $f(T, V, conc, time)$ energetics of ads	I	12

Table VII *(continued)*

System	Electrolyte	Temperature, °C	Information obtained	Method*	Reference
C_1-C_4 hydrocarbons Plat. Pt	H_2SO_4, KOH KHCO$_3$	25–65	Q, volume ads	A, J	68
C_2H_4, C_2H_2, Pt	1 N HClO$_4$	30, 60	Q as $f(V$, time)	B, D	69
C_2H_6, Pt	1 N HClO$_4$	60	Q as $f(V$, time)	B, D	70
C_2H_6, C_4H_{10}, C_8H_{18}	Pt 1 M H$_2$SO$_4$	25, 65	Q as $f(V$, time) θ as $f(V$, conc)	A, I	71
C_3H_8, Pt	13 M H$_3$PO$_4$	80, 110	Q as $f(V$, time)	A, C	72
C_2H_6, Plat. Pt	4.3 N HClO$_4$	60	Q as $f(V$, time)	B	73
C_3H_8, Pt	80 % H$_3$PO$_4$	80–140	θ, Q as $f(V$, T, time)	A, C	74
C_4H_{10}, Plat. Pt	3.7 M H$_2$SO$_4$	95	θ, Q as $f(V$, time)	B	75
C_4H_6, Pt	NaC$_2$H$_3$O$_2$		capacity as $f(V$, conc)	K	76
C_6H_6, Pt	H_2SO_4, H$_3$PO$_4$, NaOH	30–70	θ as $f(V$, conc, time, pH) energetics of ads	I	13
Naphthalene Au	HClO$_4$, NaClO$_4$	25	Γ, θ as $f(V$, conc)	I	77
Benzene naphthalene phenanthrene cyclohexane	Au 1 N H$_2$SO$_4$	25	θ as $f(V$, conc, time)	I	78
Naphthalene Ni, Fe, Cu, Pt	1 N NaClO$_4$	25	Γ or $f(V$, conc) free energy	I	21
n-decylamine, Ni, Fe, Cu, Pb, Pt	NaClO$_4$, NaOH	25	Γ as $f(V$, conc) free energy	I	22
thiourea, Ni	0.5 M Na$_2$SO$_4$	25	Γ as f(conc)	I	79
CH$_3$COONa	0.1 N Na$_2$SO$_4$		Γ, capacity as $f(V$, conc)	E, I	80
CF$_3$COOH/CF$_3$-COOK, HCOOH/HCOOK, Pt, Au, Pd	aqueous and nonaqueous	5–30	Q, θ, capacity	A, C, K	81
CH$_3$COOH/CH$_3$-COONa, Pt, Au, Ir, Ni	buffers	20	Q	A	82

from the usual adsorption studies, particularly from radiotracer measurements, and that resulting from studies of dehydrogenation, hydrogenation, etc. at open circuit.

6. THE ADSORPTION CHARACTERISTICS OF SOME SELECTED COMPOUNDS

6.1. Carbon Monoxide

Carbon monoxide chemisorbs on Pt from aqueous solutions to essentially full coverage between 0.4 and 0.9 V_{NHE}, the coverage falling off rapidly to zero at about 0.91 V. An indication of the strength of the Pt-CO bond is obtained from the fact that chemisorbed CO is not displaced by hydrogen even at potentials negative to 0.0 V_{NHE}. CO remains adsorbed at a full monolayer even after essentially all traces of CO have been removed from solution. Using transient techniques, Gilman was able to distinguish between two types of adsorbed CO species, which he termed linear or one-site, and bridged or two-site adsorption following evidence for these two species from infrared spectroscopy measurements in the gas phase. An examination of the gas phase and catalysis literature on CO adsorption should permit a better understanding of the types of bonding involved. However, with the larger amount of information available, the situation for CO adsorption in the gas phase is in perhaps a considerably more confused state. Eischens and Pliskin [84] have given a fairly complete discussion of CO adsorption studies in the gas phase using infrared spectroscopy from which the linear and bridged forms of adsorbed CO were suggested. In more recent studies of CO adsorption on Ni using infrared techniques, Blyholder [85] found strong absorption bonds at 1940 and 2080 cm^{-1} and a medium bond at 435 cm^{-1} but no other. Introduction of oxygen into the system gave several other bonds. From this he concluded that the two strong bonds arise from independent structures but that there is no evidence to assume a bridged structure. Using field emission and ion microscopes, considerably more detailed information for the adsorption of CO on W has been obtained [86]. Four distinct states of adsorbed CO were found with desorption energies of 20, 53, 75, and 100 kcal/mole. The α state (lowest desorption energy) was characterized by a spread of interaction energies, was formed on bare metal sites only in the presence of other states, and was in equilibrium with CO gas. The other three states, β, β_2, and β_3, occured on distinct crystal planes. The low index planes, particularly the (110) plane, were covered only after the other planes were already heavily covered with CO. It was suggested that all the planes were involved in bonding the energetic β states and therefore special planes on which only the α form might be held were unlikely. Ehrlich further suggested that the complexity of the surface system is such that generalizations are difficult and must be subjected to careful experimental tests. Another study which applied spectrophotometric methods proposed the presence of five different species of CO adsorbed on Ni [87] that were assigned to five observed intensity bonds. The intensities of these adsorption bonds were found to vary independently as the experimental conditions were changed [88], further suggesting the species were different. Many of the observed variations may well be explained by the surface heterogeneity of the metal, which could have sites with a rather wide range of adsorption energies. Studies of metal surfaces now being made, e.g., with low energy electron diffraction, show that the adsorption of a single gas on a single crystal surface is vastly more complex than had been previously imagined [7]. Clean metal surfaces are not static assemblies of atoms, but the surface atoms have a degree of mobility, and when a gas is strongly adsorbed the substrate atoms can move to new positions. This surface reconstruction often occurs readily at room temperature and is well known, e.g., for oxygen adsorption. Burwell and Peri [89] have concluded from their review of CO adsorption that more careful study is needed and especially independent verification of bond assignments that have been based on doubtful analogies to spectra of simple carbonyls or gaseous CO ions. The same statement that more careful study is needed is even more true of CO electrosorption where the work has been very limited and the methods are very few.

6.2. Formic Acid

The chemisorption of formic acid on Pt anodes has been studied only by coulometric methods. The information obtained from these measurements is summarized in Fig. 1. Typical "bell-shaped" behavior of the coverage–potential curves is observed for cases that have been corrected for hydrogen coverage. The rate of adsorption of HCOOH was found to be fairly slow, 150 sec being required to obtain equilibrium coverage [35]. It was concluded that formic acid is adsorbed according to a Langmuir-type isotherm.

The nature of the electrosorbed species has been the subject of much discussion, with the result that several possibilities have been

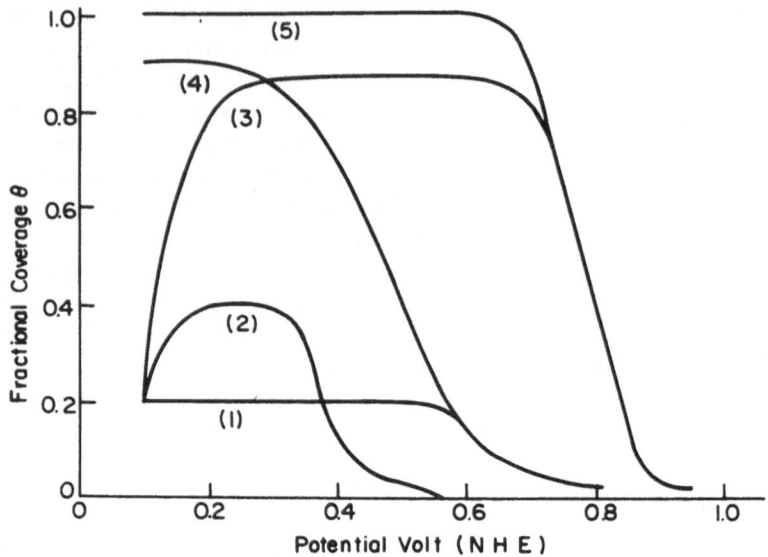

Fig. 1. Fraction of Pt surface covered with HCOOH under various experimental conditions: (1) 10^{-4} M, 30°C; (2) 1 M, 90°C; (3) 1 M, 25°C; (4) 1 M, 40°C; (5) 1 M, 30°C (from Brummer and Makrides, *J. Phys. Chem.* **68**, 1448 (1964); Breiter, *Electrochim. Acta* **8**, 447 (1963); Fleischman, Johnson, and Kuhn, *J. Electrochem. Soc.* **111**, 602 (1964).

proposed with no definite conclusion being formed. In the solution phase, evidence on the type of adsorbed species is necessarily indirect. The information, primarily obtained from the anodic oxidation of HCOOH, concerning the nature of adsorbed HCOOH is summarized in Table VIII [4]. It would appear that the formic acid molecule is favored by the available evidence for electrosorption. Considerably more work on this subject has been done in the gas phase in connection with the catalytic decomposition of HCOOH. This has recently been summarized by Bond [1], who concludes; "It is evident that there remains substantial disagreement concerning the normal adsorbed state of formic acid. There is a *prima facie* case for believing that the structures and species present on nickel are not the same at all temperatures." The absorption bands observed with infrared have been generally assigned to the form-ate ion (HCOO−) by analogy with metal formates. It appears that the formate ion is formed on Ni catalysts between 20 and 100°C. The formic acid molecule is favored at −60°C. On MgO at temperatures

Table VIII. Evidence for the Nature of Electrosorbed HCOOH

Species	Observation	Reference
HCOOH	Optimum acidity for oxidation is in the range of pH 0–1	90
HCOOH	Addition of HCOOH caused no change in cyclic polarograms in 2 N NaOH while oxidation peaks were observed in 2 N H$_2$SO$_4$	91
HCOOH	Observed catalytic decompositions for HCOOH oxidation were much larger than could be accounted for by the calculated limiting diffusion currents of other species [HCOO$^-$, (HCOOH $\cdot$ HCOO)$-$, (HCOOH)$_2$, etc.]	35
not HCOO$^-$	Methanol oxidized almost quantitatively to formate in basic solution but completely to CO$_2$ in acid	41, 90, 92
not HCOO$^-$	θ decreases with increasing anodic potential making adsorption of a negatively charged species unlikely	36
not HCOO	Maximum charge observed for adsorption from anodic charging ($Q = 210$ μCb/cm^2) would imply area of formate radical of 7 Å^2 as compared with $\sim$14 Å^2 for HCOOH	36
not CO	Q for CO at $\theta = 1$ ($\sim$340 μCb/cm^2) is higher than Q for HCOOH ($\sim$210 μCb/cm^2) and anodic traces are significantly different	36
not CO	Lack of similarity in cyclic polarograms for oxidation of CO and of HCOOH	93
not CO	CO requires much higher potential to oxidize than HCOOH	30, 34

below 100°C, the formic acid molecule has been favored [94]. From mechanism studies of HCOOH decomposition over Al$_2$O$_3$ and MgO, the formate ion is formed at 200°C [95]. From infrared spectra for HCOOH adsorption on Pt, the formate ion has been suggested as the adsorbed species [96]. On Ni it has been concluded that the adsorbed species was positively charged [97], negatively charged [1, 98], or the covalently chemisorbed neutral molecule [99]. Mars et al. [98] have suggested that differences in results may in part be due to observations on Ni-H catalysts rather than free Ni, since at lower temperatures HCOOH cannot displace adsorbed H. They conclude that more surface coverage measurements are needed, particularly with special precautions taken to prevent poisoning or contamination, and that a lack of knowledge of the precise distribution of exposed crystallographic planes and the number of sites necessary for the adsorption of a single molecule must be filled in.

A survey of mechanism studies on the anodic oxidation of formic acid indicates that a process involving the dissociative adsorption of HCOOH is favored [4]. Such studies have not been carried out in sufficient detail or under acceptable control of the system to permit reliable conclusions concerning the mechanism. The situation could be considerably clarified by establishment of the nature of the species participating in the reaction, i.e., the adsorption process.

6.3. Alcohols

Typical curves for the potential dependence of coverage obtained for the adsorption of methanol on Pt anodes determined by coulometric measurements are given in Fig. 2 [83]. The potential and concentration dependence are very similar to that generally found for the electrosorption of neutral organic molecules. It has been suggested that

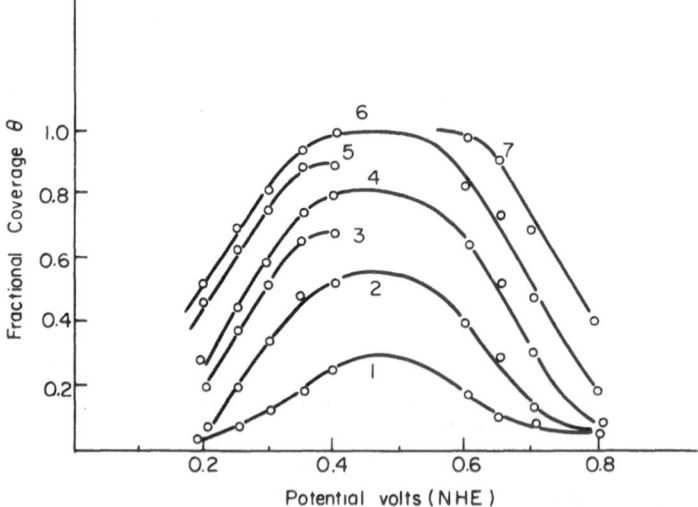

Fig. 2. Dependence of methanol adsorption on the electrode potential in 1 N H_2SO_4 and various methanol concentrations: (1) 10^{-3} M;   (2) 10^{-2} M; (3) 5×10^{-2} M;   (4) 10^{-1} M;   (5) 5×10^{-1} M;   (6) 1 M;   (7) 5 M (from Vasilyev and Bagotzky, "Fuel Cells, Their Electrochemical Kinetics," Vasilyev and Bagotzky, eds. trans. Consultants Bureau, New York, 1966).

methanol adsorption follows the Temkin isotherm and that the potential at which the methonal coverage begins to decrease coincides with the potential of oxide formation [42]. These points should receive further attention for experimental confirmation.

The principal gas phase studies with alcohols bearing on the nature of adsorption concern catalytic dehydrogenation, which is the predominant process occurring when alcohols decompose on metal catalysts. In general, it has been found that primary alcohols yield aldehydes; however, methonal commonly decomposes to CO and H_2, and secondary alcohols yield ketones [1]. Since dehydrogenation of alcohols is an endothermic process, it will be favored at increased temperatures. From an analysis of gases evolved when a platinized Pt electrode is immersed in solutions of alcohols, processes of dehydrogenation, hydrogenation, and self-hydrogenation were proposed [100]. When the Pt electrode is saturated with hydrogen, the hydrocarbon containing the same number of carbon atoms as the original alcohol predominates in the gas phase. A Pt electrode essentially free from hydrogen gives cleavage predominantly between the first carbon atoms. Decomposition of the C_1-C_2 bond proceeded more readily with aldehydes than with ethanol. It was concluded that alcohols in the presence of Pt electrodes are thermodynamically unstable substances. This point has been examined in more detail by studies of the open-circuit potential behavior by Podlovchenko [100] and will be discussed in a later section.

A few studies, particularly using infrared techniques, have been concerned with the structures of adsorbed alcohols. The interaction of ethanol with Al_2O_3 was reported to yield surface esters of the type $Al-O-C_2H_5$ [101]. The only stable surface species produced by the adsorption of methanol on silica-supported Ni at 20°C was chemisorbed CO [102]. The principal reaction of ethanol on Ni was to break the carbon—carbon bond, producing an adsorbed hydrocarbon fragment and chemisorbed CO. Some ethanol was also thought to produce a species of the structure CH_3-CH_2-Ni or CH_3-CH_2-O-Ni [102]. Blyholder and Neff noted the ready cleavage of $C-C$ and $C-H$ bonds when one of the carbons had an attached OH group. The rapid exchange of the hydroxyl hydrogen with deuterium on metal catalysts suggested that the first step in alcohol decomposition is dehydrogenation to an aldehyde which further decomposes with breakage of the $C-C$ and $C-H$ bonds. From a gravimetric study of alcohol adsorption on Al_2O_3 catalysts, the following conclusions were suggested [103]:

1. The monolayer adsorption at 25°C of lower normal aliphatic alcohols on a number of Al_2O_3 catalysts decreases in the order $CH_3OH > C_2H_5OH > n-C_3H_7OH > n-C_4H_9\dot{O}H > n-C_5H_{11}OH$.
2. At moderate temperatures, besides orientations of alcohol molecules normal to the catalyst surface (adsorption due to hydroxyl groups), planar orientation also occurs (molecule adsorbed flat on surface).
3. At higher temperatures, the molecules of all alcohols are uniformly oriented with their hydroxyl groups directed toward the catalyst surface.
4. Adsorption from gas phase and from solution are not always comparable because the presence of adsorbed solvent molecules can affect the orientation of the alcohol molecules.

No mention of the various decomposition processes of alcohols was made in this study.

6.4. Hydrocarbons

The adsorption of hydrocarbons on transition metals has been extensively studied in the gas phase with a particular emphasis on whether adsorption is associative or dissociative. These studies are of particular interest in electrosorption where fairly detailed studies of ethylene have been made. We shall attempt to summarize the results of hydrocarbon adsorption measurements in the gas phase and then compare these with adsorption at electrodes in solution [104].

The results of Twigg and Rideal [105] on the hydrogenation of C_2H_4 at Ni surfaces could be best explained if the carbon—carbon double bond broke in chemisorption, giving a complex bond by two point attachment, i.e., $NiCH_2—CH_2Ni$. Subsequent work by Conn and Twigg [106] and by Beeck et al. [107] confirmed this conclusion. Twigg and Rideal [108] were able to show that favorable spacing of the metal atoms for this mechanism was available on Ni and the other metals active in catalyzing ethylene hydrogenation. However, using Ni films, Beeck [109] found that the main process involved breakage of C—H rather than C—C bonds. Trapnell [110] concluded that initial chemisorption of C_2H_4 on W films was a four-site process but that final adsorption was a two-site process. He also found that hydrogen gas is evolved when C_2H_6 is adsorbed on several metals but that CH_4 was not extensively sorbed and no gas liberation was detected [111]. This point was confirmed by Kemball and co-workers [112]. Jenkins and Rideal [113] obtained results from self-hydrogenation of C_2H_4 on Ni that fit the equation

$$10 \; NiH + 10C_2H_4 \rightarrow 3Ni_2C_2H_4 + Ni_4C_2H_2 + 6C_2H_4$$

suggesting that associative and dissociative adsorption occur simultaneously. From the change in magnetic susceptibility of Ni powder during chemisorption of C_2H_4 at room temperature, Selwood [114] favored a two-bond adsorption. Assuming that the formation of a Ni—C bond affects the magnetic properties of Ni in the same way as a Ni—H bond, the chemisorption of C_2H_4 resulted in the Ni gaining on the average slightly more than two electrons per molecule. This implies that most of the ethylene is associatively sorbed but that a moderate fraction is held in a dissociated form requiring four or more sites. C_2H_4 on Ir was found to undergo self-hydrogenation to C_2H_6 at both 27 and 100°C, with some CH_4 also produced [115]. The degree of self-hydrogenation of C_2H_4 and of decomposition of C_2H_4 was strongly dependent on the initial amounts adsorbed.

In a study of the adsorption of hydrogen and several hydrocarbons on Ir using a field emission microscope, Arthur and Hanson [116] found that below 77°K, H_2, C_2H_2, and C_2H_4 are chemisorbed and both C_2H_2 and C_2H_4 are adsorbed without dissociation. The hydrocarbon species once chemisorbed were substantially immobile below 700°K. Ethane was largely physically adsorbed and a large portion was readily desorbed at 100°K, although residues remained which were not desorbed below 1000°K. Above 200°K, C_2H_4 decomposed on Ir to C_2H_2 and H, the H being desorbed in the temperature range 250 to 400°K, while the C_2H_2 dehydrogenated in the range 400 to 600°K to form a carbon residue. Bond [1] concluded that self-hydrogenation is absent from ethylene chemisorption at −78°C on Ni films and is of little importance over Pd at this temperature where most of the C_2H_4 is held by associative attachment, in which the π bond is broken and two carbon—metal bonds are formed. There is also considerable evidence for the associative form existing after C_2H_4 sorption on H-covered Ni-silica and on films when the resulting complex is exposed to hydrogen.

Infrared studies of olefin adsorption indicate the presence of both olefinic and paraffinic C—H bonds [99, 117]. More recent magnetic susceptibility measurements by Selwood [118] favor associative adsorption of C_2H_4 on Ni at 0°C but the dissociatively adsorbed species becomes more important at 28°C. Results of radiochemical studies [119] of ethylene chemisorption on Ni, Rh, Pd, Ir, and Pt indicate two modes for adsorption; one is involved in hydrogenation and molecular exchange, and the other is an inert form that is retained on the surface. These two modes were supposed to be associative and dissociative adsorption and better correlation of results were obtained if the retained, i.e., unreactive, species was considered to be associatively bonded. It was shown that acetylene could cause desorption of even the portion of the adsorbed ethylene layer that was retained in other experiments. Once adsorbed, the C_2H_2 behaved in a manner similar to C_2H_4 in the presence of hydrogen with regard to extent of retention on the surface. The retentions of C^{14} were further in agreement with observations that C_2H_4 adsorption on evaporated Ni, Pd, and W is accompanied by the production of C_2H_6 by self-hydrogenation. Fragmentation of the adsorbed species appears to become increasingly important as the temperature is increased. Other evidence for retention of the double bond in adsorption, i.e., associative adsorption, is obtained from studies of olefin and acetylene complexes of transition metals [120]. In olefin complexes of Ag and Pt, the olefin proton resonance (from proton resonance spectra) does not differ greatly from that in the free olefin, which is a good indication that the double bonds are retained [121].

The structure of chemisorbed acetylene has not been definitely resolved, although several studies determining the infrared spectra of the sorbed species on several metals have been carried out [122]. The spectrum observed after chemisorption of C_2H_2 on a Pd-silica catalyst showed no evidence for $C-H$ bonds of saturated hydrocarbons. Absorption bands at 3090 and 3030 cm^{-1} were attributed to $C-H$ stretching vibrations of olefinic species [117]. Eischens indicated that the preferred structure for adsorbed acetylene is of the form

$$
\begin{array}{ccc}
H & & H \\
\diagdown & & \diagup \\
C & = & C \\
| & & | \\
Pd & & Pd
\end{array}
$$

A number of studies of the electrosorption of hydrocarbons, particularly ethylene, benzene, and propane on Pt, have been carried out as indicated in Table VII. Typical results for the adsorption of ethylene on Pt as determined by the radiotracer technique are seen in Fig. 3 [12]. A Langmuir-type adsorption isotherm was observed and the rate of ethylene adsorption was controlled by diffusion to the Pt surface. The mode of adsorption was considered to be associative. Calculations of

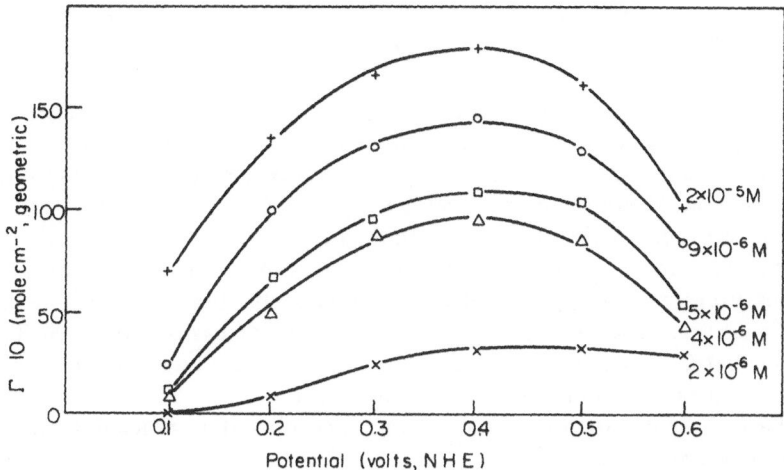

Fig. 3. Ethylene coverage as a function of potential on a platinum electrode in
$1\ N\ H_2SO_4$: $(+)\ 2 \times 10^{-5}\ M$; (m) $9 \times 10^{-6}\ M$; (m) $5 \times 10^{-6}\ M$; (m) $4 \times 10^{-6}\ M$;
$(x)\ 2 \times 10^{-6}\ M$ (from Gileadi, Rubin and Bockris, *J. Phys. Chem.* **69**, 3335 (1965).

the energetics of adsorption from gas phase data show that the associative mode is only slightly favored over the dissociative mode. Such
calculations are limited, however, since the bond energies in adsorbed
species are not likely to be the same as in the gas phase. Further support
for associative adsorption of C_2H_4 is obtained from a comparison of
the kinetics of the anodic oxidation of C_2H_4 [123] and C_2H_2 [124]. The
2.3 (RT/F) Tafel slope for C_2H_2 oxidation as compared with a 2.3
(2RT/F) Tafel slope for C_2H_4 indicates that the reactions have different
rate-limiting steps. This suggests that the reacting species are different,
which is not the case if C_2H_4 were dissociatively adsorbed.

The degree of surface coverage of Pt electrodes by C_2H_4 has been
the subject of some discussion. The first coverage measurements
reported for the adsorption of C_2H_4 on platinized gold foil electrodes
from $1\ N$ NaOH at 60°C using the radiotracer technique found a low
coverage on the order of $\theta = 0.1$ [65]. Similar values of θ were reported
for adsorption onto platinized Pt at 80°C from $1\ N\ H_2SO_4$ using coulometric techniques [66]. Refined measurements [12] with the radiotracer
method for adsorption from $1\ N\ H_2SO_4$ from 30 to 70°C onto platinized
Au gave limiting coverages in the range $\theta = 0.35$ to 0.4. These coverage
values were obtained assuming a double-layer capacity for the Pt
electrode similar to that for Hg in order to calculate surface area.

Volumetric and galvanostatic studies of C_2H_4 adsorption onto Pt black electrodes gave limiting coverages in the range of $\theta = 0.75$ to 0.77 [68]. Further results using coulometric techniques found a limiting coverage for ethylene of $\theta = 0.75$ and also indicated that on the average, 3.5 sites per C_2H_4 molecule were required for adsorption [69]. These latter results are in agreement with results from steady-state kinetic measurements that required high coverage, in the range $\theta \geqslant 0.8$ and about four sites per molecule, to correlate the experimental data with the proposed mechanism [123]. The differences between the radiotracer measurements and the coulometric results probably lie primarily in the surface area determinations, a factor of two being easily explicable. The fact that C_2H_4 appears to be adsorbed associatively on electrodes while both associative and dissociative adsorption may occur in the gas phase is again attributed to the presence of water molecules on the electrode surface.

The adsorption of several saturated hydrocarbons onto Pt black electrodes has been examined by volumetric and galvanostatic methods [68]. The results showed that hydrocarbons fall into several groups in terms of surface coverage. Methane is adsorbed to only a small extent, i.e., $\theta < 0.1$. Higher molecular weight saturated hydrocarbons adsorb to a somewhat greater extent, e.g., C_2H_6 and C_3H_8 give $\theta \simeq 0.2$ to 0.3. Unsaturated hydrocarbons give θ on the order of 0.7 to 0.8. These results are essentially in line with gas phase studies showing that saturated hydrocarbons are often only physically adsorbed. The electrolyte from which adsorption was carried out had a marked effect on the adsorption behavior of saturated hydrocarbons; however, no detectable effects were observed for unsaturated compounds. This may be understood in view of the necessary replacement of adsorbed species, which would depend on the electrolyte, and the energetics of dissociative adsorption. Further studies of ethane electrosorption have indicated that the structure of the adsorbed species consists of species between the composition of C_2H_2 radical and CO [73]. From galvanostatic studies of propane adsorption on Pt it was suggested that at 0.2 V_{NHE}, each C_3H_8 molecule covered one Pt surface atom, but from 0.3 to 0.5 V_{NHE}, three sites were involved [72]. At 0.25 V_{NHE}, the initial chemisorption appeared to be on one site but then reverted to three-site attachment, and all subsequently chemisorbed material occupied three sites. Propane was assumed to dissociate on the Pt surface. Studies to elucidate the structure of the adsorbed propane species using anodic desorption techniques suggested that three distinct adsorbed residues were present [74]. These were supposed to release 1.8, 1.3, and 6 electrons per covered Pt site when oxidized. Studies of butane adsorption on Pt black gave indications of some fragmentation of the hydrocarbon on the surface [75]. The adsorption of butane has been indicated to be diffusion controlled from radiotracer measurements [71]; however, coulometric methods have suggested that adsorption rather than mass transport was the limiting factor [75]. In general, the assignment of structure to adsorbed species from the various coulometric methods has been by indirect and, in general, in a rather obscure manner. Thus, it would seem that conclusions from these types of measurements should be taken with considerable caution. In general, it should be expected that saturated hydrocarbons would undergo dissociation and possibly fragmentation in adsorption since there is no other way to chemisorb.

The rate of benzene adsorption [13] was found to be controlled by mass transfer at low coverages as indeed was the electrosorption of most hydrocarbons studied. Benzene was found, however, not to fit either a simple or modified Langmuir adsorption isotherm, but fairly good agreement with the Temkin isotherm was obtained [13]. It was suggested that benzene adsorbs on Pt with a probable loss of aromatic character, i.e., by breaking the double bonds, and that on the average nine water molecules were replaced by a single benzene molecule. Radiotracer studies of the electrosorption of several aromatic hydrocarbons at the Au-electrolyte interface have shown that the planar, i.e., flat, orientation of the adsorbed species is preferred [78]. Naphthalene adsorbs parallel to the electrode surface on several metals, and there are indications that some Π-bonding may occur [21]. In the adsorption of n-decylamine, best agreement with experiment was obtained by assuming nonlocalized adsorption and nonrigid molecules, allowing the formation of dipole pairs [21]. The orientation of the adsorbed species was found to depend on the metal electrodes, e.g., the amine group was adsorbed toward the metal on Pt but toward the solution on Cu and Ni.

In general, one finds that where comparable systems have been studied, agreement between gas phase adsorption and electrosorption is found. There is a need for considerable experimental work before valid structure can be assigned to adsorb species both in the gas phase and at electrodes in solution. It should be noted that the tendency to dissociate and fragment in gas phase adsorption is considerably lessened in electrosorption, probably due to the presence of solvent molecules on the electrode surface. This point shall now receive further attention in reference particularly to alcohols, which have been examined by Frumkin's school using open-circuit potential measurements.

7. OPEN-CIRCUIT ADSORPTION BEHAVIOR

The subject to be discussed here is not the mechanism of establishment of open-circuit potentials in the presence of organic species, which has already been reviewed to some extent [4], but the information which such studies have given concerning the nature of the adsorption process and the adsorbed species. Mass-spectrophotometric analysis of gases produced when acetaldehyde was added to a $0.1 \, N \, H_2SO_4$ solution at a platinized Pt electrode gave $46 \, \% \, C_2H_6$ and $54 \, \% \, CH_4$, indicating hydrogenation [125] when the electrode potential was negative to $0.080 \, V_{NHE}$; at higher electrode potentials, acetaldehyde was assumed to dehydrogenate. Halide ions retarded both the dehydrogenation and hydrogenation processes. Propanol and butanol were assumed to be hydrogenated when added to a platinized Pt electrode at $0.060 \, V_{NHE}$ [125]. Analysis of the gaseous products gave for propanol $75 \, \% \, C_3H_8, C_2H_6$, and small amounts of C_2H_4, and for butanol, $76 \, \% \, C_4H_{10}$, $23 \, \% \, C_3H_8, C_2H_6, C_2H_4$, and $1 \, \% \, CH_4$. These results were found in disagreement with reported liquid-phase catalytic studies, which reported no hydrogenation of C_3H_7OH or C_4H_9OH at platinized Pt. It was concluded that the $C-O$ bond undergoes hydrogenation to the greatest extent, the C_1-C_2 bond to a much smaller extent, and there is practically no cleavage of the other $C-C$ bonds. For the hydrogenation process to occur, it was considered essential for the alcohol molecule to be oriented either with the OH group toward the metal surface or more probably for the molecule to be adsorbed flat on the surface. The slight maximum before the beginning of the potential plateau for galvanostatic charging curves of acetaldehyde adsorbed at a Pt electrode from $0.1 \, N \, H_2SO_4$ were explained by polymerization of the CH_3CHO, which required additional energy for oxidation to begin [126].

Podlovchenko and Gorgonova suggested that their experimental data on methanol adsorption on Pt obtained from charging curves was in good agreement with the hypothesis that dehydrogenation products of CH_3OH are chemisorbed on the platinized Pt surface [45]. It was also suggested that possibly a stronger chemisorption is characteristic of species containing oxygen. From a detailed analysis of the data, it was supposed that methanol molecules primarily yield three hydrogen atoms on adsorption and a particle of the type $O-C-H$ was left chemisorbed on the Pt. In this treatment it was assumed that the chemisorbed species occupied just as many sites as it gives up electrons upon oxidation. These authors also stated that in spite of the possibility of obtaining specific information with the anode pulse method, e.g., as used by Breiter and Gilman, it cannot be stated with complete assurance that their adsorption measurements pertain to adsorbed CH_3OH and so do not include to a significant degree the chemisorbed products of its oxidation.

The introduction of propyl alcohol to a platinized Pt electrode at $0.064 \, V_{NHE}$ resulted in the evolution of gases which consisted of $19 \% \, C_2H_6$, $76 \% \, C_3H_8$, and $4 \% \, C_4H_{10}$ [127]. Gases evolved at platinized Pt at 0.041 V consisted of $< 1 \% \, CH_4$, $< 1 \% \, C_2H_6$, $23 \% \, C_3H_8$, and $76 \% \, C_4H_{10}$; while at 0.50 V, analysis of the gases gave $2 \% \, C_2H_6$, $69 \% \, C_3H_8$, and $28 \% \, C_4H_{10}$. The final products were strongly dependent on the potential at which the alcohol was introduced into the system. It was observed that dehydrogenated species could have rather large effects on electrode processes. For example, the formation of residues poisons the electrode and prevents the oxidation of alcohols; at $0.40 \, V_{NHE}$, the rate of oxidation of alcohols decreased by a factor of 10^4 from processes occuring on clean electrodes. By adsorbing molecules, removing the electrode and washing it, and then oxidizing the material remaining on the electrode, the nature of the adsorbed species was examined. Tafel lines obtained in methanol solution were compared with Tafel lines obtained by oxidizing the residue remaining after electrode washing, and a difference in overvoltage of only about 40 mV was found [127]. From this it was concluded that the oxidation of methanol was limited by the rate of removal of the firmly bound species. Larger overvoltage differences were found with other alcohols; and with HCOOH, the difference was such that it could not be assumed that the firmly bound residue was an intermediate.

When alcohols were brought in contact with a degassed platinized Pt electrode, a cleavage of the C_1-C_2 bonds was predominant [100]. A decomposition of the C_1-C_2 bond was found to proceed much more readily for acetaldehyde than for ethanol. This parallels a suggestion from gas phase studies [102] that aldehydes are more easily decomposed than alcohols. For ethanol, only about half of the chemisorbed species could be displaced by interaction with hydrogen. The amounts of chemisorbed ethanol and hydrogen were strongly dependent on the ethanol concentration; however, the final values of the open-circuit potentials were essentially independent of concentration and were assumed to be determined by hydrogen. For a number of substances, the final value of the open-circuit potential was independent of the potential at which the substance was introduced into the system; however, aldehydes were an exception. It was proposed that the steady states of the surface for several substances differ considerably when they are introduced at low and high values of potential, e.g., as a result of chemisorption of various oxidation products.

The behavior of methanol at Pt electrodes differed considerably from that of other saturated alcohols, presumably due to the absence of the $C-C$ group [46]. At low potentials, near zero on the hydrogen scale, no gas was evolved from methanol adsorption, and chemisorbed species did not accumulate with time.

In summary of this work, it appears that when a steady potential is established at platinized Pt electrodes in solutions of alcohols and aldehydes containing more than one carbon atom, processes of dehydrogenation and self-hydrogenation of the original substances and their decomposition products (mainly along the C_1-C_2 bond), occur [55]. A steady concentration of H_{ads} on the electrode surface, established and maintained by these processes, determines the final open-circuit potential.

Several suggestions concerning the nature of the species resulting from methanol and formic acid have been made and discussed. Many investigators have accepted the proposal by Giner [128] of a "reduced CO_2" species having the form of carbon monoxide or formate radical. Investigations by Vielstich [93] have shown conclusively that the species resulting from HCOOH adsorption is not CO. More recent investigations of the nature of the interaction of CO_2 with adsorbed H have shown that the assumption of CO_2 reduction is not necessary and that the results could be better explained by simply poisoning of the H oxidation by interaction with CO_2 [129]. It may be suggested that this type of process, i.e., interaction of CO_2 resulting from oxidation of the organic with hydrogen present on the electrode, may be a general poisoning species below 0.3 V_{NHE}.

8. SUMMARY

In the past, electrosorption has contributed little to the general knowledge of adsorption phenomena and has relied heavily on gas phase studies. While there are many techniques available for the study of gas phase adsorption that cannot be used when solvent is present, they have not been used with great success to determine the nature of adsorbed species at the electrode surface. It would be of great value to have more detailed studies of systems in the gas phase that could then be compared with solution phase adsorption at electrodes. The addition of water vapor in gas phase studies would greatly aid in making comparisons with solution phase measurements, since then both processes would be competitive or replacement reactions. The presence of solvent molecules is the one large difference in electrosorption measurements, and this is strongly in evidence in the energetic aspects of adsorption. The ability to vary the energy of the systems much more readily at electrodes, i.e., by control of the metal-solution potential difference, is a considerable advantage for electrosorption.

In general, it can be stated that the surface structure of adsorbed organic species has not been established for any case in electrosorption but also has not been established for any case in the gas phase. Considerable progress has been made in gas phase study by the introduction of ultrahigh vacuum techniques with a strong emphasis on system purity and control. While such techniques are not applicable to electrosorption, the control of system and purity are extremely important, and improved techniques and procedures are now becoming available. Finally, it should be stated that gas phase catalysis and electrosorption are not unrelated since both are aspects of adsorption and each will have much to contribute to the other if high standards of quality in experimental measurement are required and maintained.

REFERENCES

1. Bond, *Catalysis by Metals*, Academic Press, New York (1962).
2. Hayward and Trapnell, *Chemisorption*, 2nd ed., Butterworth, Washington, D. C. (1964).
3. Frumkin and Damaskin, *Modern Aspects of Electrochemistry*, Vol. 3, Chap. 3, Bockris and Conway, eds., Butterworth, Washington, D. C. (1964).
4. Piersma and Gileadi, *Modern Aspects of Electrochemistry*, Vol. 4, Chap. 2, Bockris, ed., Plenum Press, New York (1966).
5. Gileadi and Conway, *Modern Aspects of Electrochemistry*, Vol. 3, Chap. 5, Bockris and Conway, eds., Butterworth, Washington, D. C. (1964).
6. Ehrlich, *Advan. Catalysis* **14**, 255 (1963).
7. May, *Ind. Eng. Chem.* **57**, 19 (1965).
8. Roberts, *Some Problems in Adsorption*, Cambridge University Press, London (1939).
9. Schuldiner and Warner, *J. Phys. Chem.* **68**, 1223 (1964); *Electrochim. Acta.* **11**, 307 (1966).

10. Schuldiner, Piersma, and Warner, *J. Electrochem. Soc.* **113**, 573 (1966).
11. American Oil Company, Whiting Laboratories, *Quarterly Reports*, 1–9, Contract No. DA-11-022-ORD-4023.
12. Gileadi, Rubin, and Bockris, *J. Phys. Chem.* **69**, 3335 (1965).
13. Heiland, Gileadi, and Bockris, *J. Phys. Chem.* **70**, 1207 (1966).
14. Breiter, *Electrochim. Acta* **7**, 25 (1962); Bold and Breiter, *Z. Elektrochem.* **64**, 897 (1960); Breiter and Kennel, *Z. Elektrochem.* **64**, 1180 (1960).
15. Gileadi, *J. Electroanal. Chem.* **11**, 137 (1966).
16. Frumkin, *Advances in Electrochemistry and Electrochemical Engineering*, Vol. 3, Delahay, ed., Interscience, New York (1963).
17. Warner and Schuldiner, *J. Electrochem. Soc.* **112**, 853 (1965).
18. Blomgren, Bockris, and Jesch, *J. Phys. Chem.* **65**, 2000 (1961).
19. Barradas and Conway, *Electrochim. Acta* **5**, 319, 349 (1961).
20. Bockris and Swinkels, *J. Electrochem. Soc.* **111**, 736 (1964).
21. Bockris, Green, and Swinkels, *J. Electrochem. Soc.* **111**, 743 (1964).
22. Dahms and Green, *J. Electrochem. Soc.* **110**, 1075 (1963).
23. Bewig and Zisman, *Advan. Chem. Ser.*, No. 33, 100 (1961); Bewig, Timmons, and Zisman, *NRL Report* 6200, February 12, 1965.
24. Smith and Burwell, Jr., *J. Am. Chem. Soc.* **84**, 925 (1962).
25. Bond and Wells, *Advan. Catalysis* **15**, 91 (1964).
26. Bockris, Devanathan, and Müller, *Proc. Roy. Soc.* **274**, 55 (1963).
27. Devanathan and Tilak, *Chem. Rev.* **65**, 635 (1965).
28. Frumkin, *Dokl. Akad. Nauk. Uz. SSR* **154**, 1432 (1964); *Elektrokhim.* **1**, 394 (1965).
29. Piersma, Ph.D. Dissertation, University of Pennsylvania (1965).
30. Warner and Schuldiner, *J. Electrochem. Soc.* **111**, 992 (1964).
31. Fasman, Padyukov, and Sokol'skii, *Dokl. Akad. Nauk Uz. SS* **150**, 856 (1963).
32. Sklyarov and Kolotyrkin, *Elektrokhim.* **1**, 360 (1965).
33. Gilman, *J. Phys. Chem.* **66**, 2657 (1962); **67**, 78, 1898 (1963).
34. Brummer and Ford, *J. Phys. Chem.* **69**, 1355 (1965); Brummer, *J. Phys. Chem.* **69**, 1363 (1965).
35. Breiter, *Electrochim. Acta* **8**, 447, 457 (1963).
36. Brummer and Makrides, *J. Phys. Chem.* **68**, 1448 (1964).
37. Fleischmann, Johnson, and Kuhn, *J. Electrochem. Soc.* **111**, 602 (1964).
38. Brummer, *J. Phys. Chem.* **69**, 562 (1965).
40. Breiter, *Electrochim. Acta* **10**, 503 (1965).
41. Pavela, *Ann. Acad. Sci. Fennicae, Ser. A II*, Chem. No. 59, 1 (1954).
42. Breiter and Gilman, *J. Electrochem. Soc.* **109**, 622, 1099 (1962); Breiter, *Electrochim. Acta* **7**, 533 (1962).
43. Breiter, *J. Electrochem. Soc.* **110**, 449 (1963).
44. Breiter, *Electrochim. Acta* **8**, 973 (1963).
45. Podlovchenko and Gorgonova, *Dokl. Akad. Nauk Uz. SS* **156**, 673 (1964).
46. Podlovchenko, Petrii, and Gorgonova, *Elektrokhim.* **1**, 182 (1965).
47. Khira Lal, Petrii, and Podlovchenko, *Elektrokhim.* **1**, 316 (1965).
48. Bogdanovskii, Kononovich, and Khomchenko, *Russ. J. Phys. Chem.* **38**, 1362 (1964).
49. Khazova, Vasilyev, and Bagotzky, *Elektrokhim.* **1**, 84, 439 (1965).
50. Petrii, Podlovchenko, Frumkin, and Khira Lal, *J. Electroanal. Chem.* **10**, 253 (1965).

51. Khira Lal, Petrii and Podlovchenko, *Dokl. Akad. Nauk Uz. SSR* **158**, 1416 (1964).
52. Bagotzky and Vasilyev, *Elektrochim. Acta* **9**, 869 (1964).
53. Petrii, *Dokl. Akad. Nauk. Uz. SS* **160**, 871 (1965).
54. Rightmire, Rowland, Boos, and Beals, *J. Electrochem. Soc.* **111**, 242 (1964).
55. Podlovchenko, Petrii, Frumkin, and Khira Lal, *J. Electroanal. Chem.* **11**, 12 (1966).
56. Schmid and Hackerman, *J. Electrochem. So* . **110**, 440 (1963).
57. Shashkina and Kulakova, *Russ. J. Phys. Chem.* **35**, 908 (1961).
58. Pochekaeva, *Russ. J. Phys. Chem.* **35**, 787 (1961).
59. Tsintsevich, Khomchenko, and Vovchenko, *Russ. J. Phys. Chem.* **38**, 1248 (1964).
60. Reid and Kruger, *Nature* **203**, 402 (1964).
61. Conway, Barradas, and Zawidzki, *J. Phys. Chem.* **62**, 676 (1958).
62. Newmiller and Pontius, *J. Phys. Chem.* **64**, 584 (1960).
63. Wroblowa and Green, *Electrochim. Acta* **8**, 679 (1963).
64. Breiter, *J. Electrochem. Soc.* **109**, 42 (1962).
65. Dahms, Green, and Weber, *Nature* **196**, 1310 (1962).
66. Griffith and Rhodes, *Fuel Cells*, 32 (1963).
67. Burshtein, Tyurin, and Pshenichnikov, *Dokl. Akad. Nauk Uz. SSR* **160**, 629 (1965).
68. Niedrach, *J. Electrochem. Soc.* **111**, 1309 (1964).
69. Gilman, *Trans. Faraday Soc.* **62**, 466, 481 (1966).
70. Gilman, *Trans. Faraday Soc.* **61**, 2546, 2561 (1965).
71. Flannery, Aromowitz, and Walker, American Oil Co. Reports, *No.* 1, March, 1965; *Final Report*, July, 1965, Contract DA-49-186-AMC-167 (X).
72. Brummer, Ford, and Turner, *J. Phys. Chem.* **69**, 3424 (1965).
73. Niedrach, Gilman, and Weinstock, *J. Electrochem. Soc.* **112**, 1161 (1965).
74. Brummer and Giner, Tyco Laboratories, *Fourth Intern Technical Report*, April–October, 1965, Contract DA-44-009 AMC 410 (7).
75. Shropshire and Horowitz, *J. Electrochem. Soc.* **113**, 490 (1966).
76. Mirkind, Fioshin, and Romanov, *Russ. J. Phys. Chem.* **38**, 1201 (1964).
77. Green and Dahms, *J. Electrochem. Soc.* **110**, 466 (1963).
78. Dahms and Green, *J. Electrochem. Soc.* **110**, 1075 (1963).
79. Green and Swinkels, and Bockris, *Rev. Sci. Instr.* **33**, 18 (1962).
80. Girina, Fioshin, and Kazarinov, *Elektrokhim.* **1**, 478 (1965).
81. Conway and Dzieciuch, *Can. J. Chem.* **41**, 21, 38, 55 (1963).
82. Dichinson and Wynne-Jones, *Trans. Faraday Soc.* **58**, 382, 388, 400 (1962).
83. Vasilyev and Bagotzky, *Fuel Cells, Their Electrochemical Kinetics*, Vasilyev and Bagotzky, eds., trans. Consultants Bureau, New York, p. 99 (1966).
84. Eischens and Pliskin, *Advan. Catalysis* **10**, 1 (1958).
85. Blyholder. *Proceedings of the Third International Congress on Catalysis*, Vol. 1, Amsterdam, July, 1964, Sachtler, Schiut, and Zwitering, eds., John Wiley & Sons, New York, p. 657 (1965).
86. Ehrlich, *ibid.*, p. 113.
87. Leftin and Hobson, Jr., *Advan. Catalysis* **14**, 115 (1963).
88. Yates, Jr., and Garland, *J. Phys. Chem.* **65**, 617 (1961).
89. Burwell Jr., and Peri, *Ann. Rev. Phys. Chem.* **15**, 131 (1964).
90. Buck and Griffith, *J. Electrochem. Soc.* **109**, 1005 (1962).
91. Liang and Franklin, *Electrochim. Acta* **9**, 517 (1964).

92. Schlatter, American Chemical Society National Meeting, Chicago (1961); New York (1963); *Fuel Cells*, Vol. 2, Young, ed., Reinhold, New York, p. 90 (1963).
93. Vielstich and Vogel, *Ber. Bunsengesellschaft* **68**, 688 (1964).
94. Kubokawa and Miyata, *Proceedings of the Third International Congress on Catalysis*, Vol. 2, Amsterdam, July 1964, Sachtler, Schmit, and Zwietering, eds., John Wiley & Sons, New York, p. 871 (1965).
95. Scholten, Mars, Menon, and Von Hardeveld, *ibid.*, p. 881.
96. Hirota, Kumata, and Asai, *Nippon Kagaku Zasshi* **80**, 701 (1959).
97. Schwab, *Discussions Faraday Soc.* **8**, 166 (1950); Dawden and Reynolds, *ibid.*, p. 184.
98. Mars, Scholten, and Zwietering, *Advan. Catalysis* **14**, 35 (1963).
99. Eischens, Francis, and Pliskin, *J. Phys. Chem.* **60**, 194 (1956); Eischens and Pliskin, *Proceedings of the Second International Congress on Catalysis*, Vol. 1, Editions Technip, Paris, p. 789 (1961).
100. Padlovchenko, *Elektrokhim.* **1**, 101 (1965).
101. Boreskov, Shchekochikhin, Makarov, and Filimonov, *Dokl. Akad. Nauk. Uz. SSR* **156**, 901 (1964).
102. Blyholder and Neff, *J. Catalysis* **2**, 138 (1963).
103. Vasserberg, Balandin, and Maksimova, *Russ. J. Phys. Chem.* **35**, 419 (1961).
104. The author wishes to acknowledge use of a summary of ethylene adsorption prepared by Mr. Rubin.
105. Twigg and Rideal, *Proc. Roy. Soc. (London)* **A171**, 55 (1939).
106. Conn and Twigg, *Proc. Roy. Soc. (London)* **A171**, 70 (1939).
107. Beeck, Smith, and Wheeler, *Proc. Roy. Soc. (London)* **A177**, 62 (1940).
108. Twigg and Rideal, *Trans. Faraday Soc.* **36**, 533 (1940).
109. Beeck, *Discussions Faraday Soc.* **8**, 118 (1958); *Rev. Mod. Phys.* **17**, 61 (1945).
110. Trapnell, *Trans. Faraday Soc.* **48**, 160 (1952).
111. Trapnell, *ibid.*, **52**, 1618 (1956).
112. Wright, Ashmore, and Kemball, *Trans. Faraday So* . **54**, 1692 (1958).
113. Jenkins and Rideal, *J. Chem. Soc.* **2490**, 2496 (1955).
114. Selwood, *J. Am. Chem. Soc.* **79**, 3346 (1957).
115. Roberts, *J. Phys. Chem.* **67**, 2035 (1963).
116. Arthur, Jr., and Hansen, *J. Chem. Phys.* **36**, 2062 (1962).
117. Little, Sheppard, and Yates, *Proc. Roy. Soc.* **A259**, 242 (1960).
118. Selwood, *J. Am. Chem. Soc.* **83**, 2853 (1961).
119. Cormack, Thomson, and Webb, *J. Catalysis* **5**, 224 (1966).
120. Bennett, *Chem. Rev.* **62**, 611 (1962).
121. Powell and Sheppard, *J. Chem. Soc.* 2519 (1960).
122. Eischens, *Science* **146**, 486 (1964).
123. Wroblowa, Piersma, and Bockris, *J. Electroanal. Chem.* **6**, 401 (1963).
124. Johnson, Wroblowa, and Bockris, *J. Electrochem. Soc.* **111**, 864 (1964).
125. Podlovchenko, Petrii, and Frumkin, *Dokl. Akad. Nauk. Uz. SSR* **153**, 379 (1963).
126. Podlovchenko and Iofa, *Russ. J. Phys. Chem.* **38**, 112 (1964).
127. Frumkin, *Batteries 2 Research and Development in Non-Mechanical Electrical Power Sources*, Collins, ed., Symposium Publication Division, Pergamon Press, New York (1965).
128. Giner, *Electrochim. Acta* **8**, 857 (1963); **9**, 63 (1964).
129. Piersma, Warner, and Schuldiner, *J. Electrochem. Soc.* **113** (1966).

Chapter 3

Kinetics of Diffusion-Controlled Electrosorption of Neutral Molecules

A. K. N. Reddy*

1. INTRODUCTION

Nowadays it is a foregone conclusion that measurements of current and potential alone will not lead to the unequivocal elucidation of the mechanism of electrode reactions.

This realization has stimulated an increasing interest in the concentration of species adsorbed on an electrode. For example, in the study of the kinetics of the hydrogen evolution reaction, the fraction of the surface covered with adsorbed hydrogen atoms has been studied by several methods [1]. The adsorbed species in hydrogen evolution is an *intermediate* in the reaction. This need not always be the case. Even without participating in the charge-transfer reaction, the adsorbate may nevertheless affect the rate of the reaction. A simple way in which the adsorbate may achieve this effect is by creating an occupancy problem. Thus, the adsorbed species, by covering a fraction θ of the electrode surface, permits only a fraction $(1 - \theta)$ to be available for the electrode reaction. The adsorption of neutral molecules may also alter the surface or χ-potential in the double layer, and this affects the field utilized by the electrode reaction [2, 3]. Some examples of the effect of neutral molecules on electrode reactions can be mentioned. Thus, the adsorption of alkaloids has been reported [4] to affect the magnitude of the exchange current density for the discharge of hydrogen ions on mercury and lead electrodes. Again, the effect of surface-active substances in the

* Present address: Department of Inorganic and Physical Chemistry, Indian Institute of Science, Bangalore-12, India.

phenomena of bright plating and leveling is well known in the field of electrodeposition [5].

The overall process of the adsorption of neutral molecules on an electrode may be considered to consist of two steps: (1) mass transfer from the bulk of the solution and (2) adsorption onto the electrode. The adsorption step on to an electrode site may be formally treated as a reaction in the following manner:

$$\square + A \rightarrow \boxed{A} \tag{1}$$

where A represents the adsorbate in solution at a position corresponding to the initial state of the adsorption "reaction," i.e., just adjacent to the electrode, and $\square$ is a site on the metal surface corresponding to the final state. All positions on the electrode surface can be assumed to be equivalent, i.e., there is translational symmetry in the YZ-plane parallel to the electrode. Hence, the initial and final states are any pair of opposite positions on the $x = 0$ solution plane and the $x = x_E$ electrode plane respectively (Fig. 1), and the two steps of the overall adsorption process are: (1) mass transport of the adsorbate from $x \rightarrow \infty$ to $x = 0$ and (2) adsorption, i.e., transfer from positions on the $x = 0$ plane to $\square$-sites on the electrode ($x = x_E$ plane).

Most of the studies on the influence of the adsorption of neutral surface-active substance on electrode reactions have assumed that mass transport processes do not limit the adsorption rate. This, of course, implies that the rate of adsorption is controlled by the adsorption step. In recent years, however, the influence of mass-transfer control on the

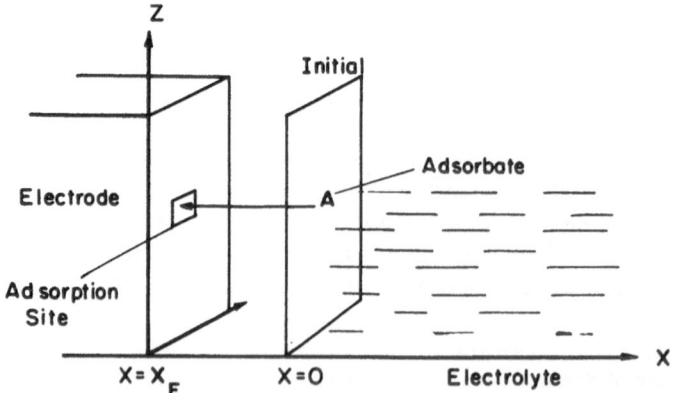

Fig. 1. Initial ($x = 0$) and final ($x = x_E$) states for adsorption.

adsorption rate has been considered both theoretically and experimentally [6-9]. Rate control by bulk transport implies, of course, that the adsorption "reaction" is at equilibrium.

It is easy to imagine that the mass transfer-controlled adsorption of neutral molecules may perhaps play an important part in the explanation of long-time effects (minutes to hours) that have been observed in several electrode reactions. Mass transport-controlled adsorption would also be of significance in situations involving a constantly renewed surface, e.g., a dropping mercury electrode.

2. THE ADSORPTION ISOTHERM AND THE MASS TRANSFER PROCESS

The method of analysis of the mass transport-controlled adsorption of electroinactive neutral molecules will now be discussed.

In the initial condition ($t = 0$), one is dealing with the *unperturbed* system, i.e., the concentration c' of the neutral adsorbable species is equal to the bulk value c^0 everywhere in the system:

$$c'(x, 0) = c^0 \tag{2}$$

If, instead of discussing the actual concentration $c'(x, t)$ it is intended only to describe the *perturbation* in the concentration, one can define the perturbation (or departure) from the bulk value thus:

$$
\begin{aligned}
c(x, t) = &\text{ (initial concentration} - \text{actual concentration} \\
&\text{ at } x \text{ and } t) \\
= &\ c^0 - c'(x, t)
\end{aligned}
\tag{3}
$$

In terms of the perturbation in concentration, $c(x, t)$, the initial condition of the system is

$$c(x, 0) = 0 \tag{4}$$

If the perturbation is zero everywhere (at any x), a concentration gradient (the driving "force" for diffusion) is absent, and there can be no nonequilibrium process of diffusion, i.e., the system is at equilibrium.

To produce diffusion, the system must be perturbed from its equilibrium state characterized by a uniform concentration everywhere. The general procedure is to remove or add material either inside the system or, more conveniently, at its boundaries. Then, there is a perturbation in concentration, $c \neq 0$, at the sink or source that has been created. This sets up a concentration gradient across the plane where the sink or source is set up, and diffusion occurs. The condition of flux continuity operates so that the number of moles of material removed or added per square centimeter per second is equal to the diffusion flux due to the concentration gradient.

The standard method of producing a perturbation is to impose a programmed current or potential across the electrode–solution interface. This leads to electrode reactions that either consume or generate material, and thus produce concentration gradients and, therefore, diffusion. The important point is that in this case there is *external* control or programming of the perturbation and, thus, of the diffusion process.

However, when there is adsorption not followed by charge transfer, one has no external control of the perturbation of the system. Instead, the rate of the adsorption "reaction" involving the neutral molecules decides the rate of consumption of these molecules at the $x = 0$-plane. The system is self-programming with respect to the diffusion process.

Using the condition of equality of fluxes at the $x = 0$ plane, one has:

$$\frac{d\Gamma(t)}{dt} = J_D \tag{5}$$

where $\Gamma(t)$ is the surface *concentration** of neutral molecules on the electrode at the time t, and J_D is the diffusion flux at $x = 0$, i.e.,

$$J_D = -D \left(\frac{\partial c}{\partial x}\right)_{x=0} \tag{6}$$

The surface concentration $\Gamma(t)$ is related to the volume concentration $c'(0, t)$ at the $x = 0$-plane by means of an adsorption isotherm.

A general form of the isotherm results from the application of the law of mass action to the adsorption reaction—equation (1). Thus,

$$\frac{a_{\boxed{A}}(t)}{a_{\square} a_A(0, t)} = k = \exp\left(\frac{-\Delta G^0}{RT}\right) \tag{7}$$

where $a_{\boxed{A}}(t)$ is the activity of the adsorbed species at the time t and is related to $\Gamma(t)$; the quantities $a_{\square}$, $a_A(0, t)$, and ΔG^0 are the activity of

* The use of the symbol Γ should not be taken to mean that one is referring to the surface *excess*. The latter quantity is the *integral* of the perturbation in concentration with respect to distance from a reference plane, e.g., the electrode, and becomes equal to the amount adsorbed only when the bulk concentration of adsorbate is negligible.

empty sites on the metal, the activity of adsorbate at the $x = 0$-plane, and the standard free energy of adsorption, respectively. The standard free energy of adsorption, ΔG^0, is the *constant* value assumed by the free energy when the various activities in equation (7) take certain arbitrarily chosen standard values. If, however, concentrations are used instead of activities, then the variation of activity coefficients with concentration make the apparent standard free energy of adsorption ΔG^0 (based on standard concentrations) a *function* of the activity of the adsorbed species, $a_{\boxed{A}}$. It is to be noted that the use of the law of mass action implies that the adsorption step is very fast compared to mass transport, i.e., there is adsorption equilibrium.

A special case of the general adsorption isotherm results from ΔG^0 being independent of the extent of adsorption, i.e., it is a constant. One obtains, after replacing activities with concentrations,

$$\frac{\Gamma(t)}{\Gamma_\Box c'_A} = k \tag{8}$$

Defining the coverage $\theta(t)$ by the following relation

$$\theta(t) = \frac{\Gamma(t)}{\Gamma_{\max}} \tag{9}$$

where $\Gamma_{\max}$ is the maximum surface concentration, equation (8) becomes:

$$\frac{\theta(t)}{1 - \theta(t)} = kc'(0, t) \tag{10}$$

This is the Langmuir adsorption isotherm.

The adsorption isotherm, therefore, provides a link between $\Gamma(t)$, the surface concentration of adsorbed species, and the actual volume concentration $c'(0, t)$ at the $x = 0$-plane. By controlling $\Gamma(t)$, the isotherm controls $[d\Gamma(t)]/dt$, J_D, and thus $c(0, t)$. But $c(0, t)$ in turn affects $\Gamma(t)$, and so on. In this manner, the isotherm programs the perturbation of the system from concentration equilibrium.

3. ANALYSIS OF THE DIFFUSION PROBLEM

One seeks in such situations to know the response of the system to a given perturbation, i.e., to know the functional relationship between the departure $c(x, t)$ of the concentration from its initial value and the stimulus, viz., flux J_D at the boundary.

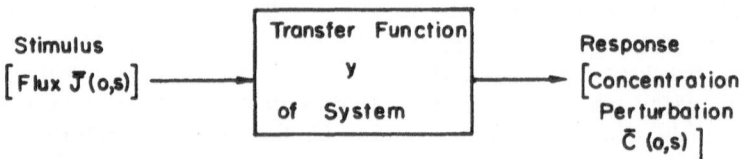

Fig. 2. Black-box approach relating flux to concentration perturbation.

A black-box approach can be adopted. The structure of the system can be ignored, and its characteristics can be represented by a *mathematical transfer function* that governs the transformation of the stimulus into response (Fig. 2). Thus we can write

$$c(0, t) = XJ(0, t) \tag{11}$$

or one can relate the logarithms of the flux and concentration. It is, however, most convenient to relate the Laplace transforms* of the perturbing flux and the concentration. Thus one writes (Fig. 2)

$$\bar{c}(0, s) = Y\bar{J}(0, s) \tag{12}$$

where Y is the yet-unspecified transfer (impedance) function characterizing the system and relating cause (stimulus J) and effect (response c); $\bar{c}(0, s)$ and $\bar{J}(0, s)$ are the Laplace transforms of $c(0, t)$ and $J(0, t)$, respectively.

In order to evaluate Y, the special case of the concentration-response of the system to the application of a constant unit stimulus at the time $t = 0$ is considered. Denoting the unit-step flux by J_{US}, one has[†]

$$\bar{J}_{US} = \int_0^\infty J_{US}\, e^{-st}\, dt = \int_0^\infty e^{-st}\, dt = \frac{1}{s} \tag{13}$$

Hence,

$$\bar{c}_{US}(0, s) = Y\bar{J}_{US} = Y\frac{1}{s} \tag{14}$$

---

* The Laplace transform $L[f(t)]$ of any function of time $f(t)$ is defined thus:

$$L[f(t)] = \int_0^\infty f(t)\, e^{-st}\, dt = \bar{f}(s)$$

where s is a complex number.

[†] The Laplace transform of a constant m is m/s, as is obvious by integrating

$$\int_0^\infty m e^{-st}\, dt$$

or

$$Y = s\bar{c}_{US}(0, s) \tag{15}$$

Introducing this expression for Y into equation (12), one gets

$$\bar{c}(0, s) = s\bar{c}_{US}(0, s) \, \bar{J}(0, s) \tag{16}$$

Now, applying the Laplace transformation to both sides of the flux continuity equation (5), one has,* with L symbolizing the Laplace transform operator,

$$L\left[\frac{d\Gamma(t)}{dt}\right] = \bar{J}(0, s) \tag{17}$$

or

$$s\bar{\Gamma}(s) - \Gamma(t = 0) = \bar{J}(0, s) \tag{18}$$

Since in all the following treatment it is assumed† that at $t = 0$ the electrode is bare of adsorbate, i.e. $\Gamma(t = 0) = 0$, equation (18) reduces to

$$s\bar{\Gamma}(s) = \bar{J}(0, s) \tag{19}$$

and, therefore, by introducing (19) in (16),

$$\bar{c}(0, s) = s^2\bar{c}_{US}(0, s) \, \bar{\Gamma}(s) \tag{20}$$

The quantity $\bar{c}_{US}(0, s)$ is obtained by a solution of the unit constant flux diffusion problem. One considers the diffusion of the species to the $x = 0$ plane given by the differential equation (Fick's second law)

$$\frac{\partial c}{\partial t} = D \frac{\partial^2 c}{\partial x^2} \tag{21}$$

which after Laplace transformation leads to the total differential equation

$$\frac{d^2\bar{c}(x, s)}{dx^2} - \frac{s}{D}\,\bar{c}(x, s) - \frac{1}{D}\,c(t = 0) = 0 \tag{22}$$

---

* The Laplace transform of the function $\{d[f(t)]\}/dt$ is $s\bar{f}(s) - f(t = 0)$, where $f(t = 0)$ is the value of the function $f(t)$ when $t = 0$.
† This assumption is only for mathematical convenience. One can proceed without assuming that $\Gamma(t = 0) = 0$—the only thing is that the mathematics becomes more cumbersome.

which becomes, with the condition that the perturbation in concentration is initially zero, i.e., $c(t = 0) = 0$,

$$\frac{d^2\bar{c}(x, s)}{dx^2} - \frac{s}{D}\,\bar{c}(x, s) = 0 \tag{23}$$

The solution of this may be written as

$$\bar{c}(x, s) = A\,\exp\left[-\,(s/D)^{1/2}x\right] + B\,\exp[(s/D)^{1/2}x] \tag{24}$$

Now, using the boundary condition $c(\infty, t) = 0$, i.e., there is no perturbation at a distance infinitely far from the source of the perturbation, the constant B is easily seen to be zero, leaving

$$\bar{c}(x, s) = A\,\exp\left[-\,(s/D)^{1/2}x\right] \tag{25}$$

To evaluate the constant A, the unit constant-flux J_{US} is linked to the concentration gradient by Fick's first law:

$$1 = J_{US} = -D\left[\frac{\partial c(0, t)}{\partial x}\right]_{x=0} \tag{26}$$

or

$$\bar{J}_{US} = \frac{1}{s} = -D\,\frac{d\bar{c}(0, s)}{dx} \tag{27}$$

Equation (25) is differentiated and combined with equation (27) to obtain

$$\left[\frac{d\bar{c}(0, s)}{dx}\right]_{x=0} = \frac{As^{1/2}}{D^{1/2}}\exp[-(s/D)^{1/2}x] = -\frac{1}{sD} \tag{28}$$

Hence

$$A = D^{-1/2}\,s^{-3/2} \tag{29}$$

Inserting this expression into equation (25) and then setting $x = 0$, it is obvious that

$$\bar{c}_{US}(0, s) = D^{-1/2}\,s^{-3/2} \tag{30}$$

This result can be combined with equation (20) to finally give

$$\bar{c}(0, s) = s^2 D^{-1/2} s^{-3/2}\,\bar{I}(s) = (s/D)^{1/2}\,\bar{I}(s) \tag{31}$$

4. THE TIME VARIATION OF THE SURFACE CONCENTRATION OF ADSORBED SPECIES

At this stage the adsorption isotherm is used to substitute for $\bar{\Gamma}(s)$. A general isotherm in which ΔG^0 is a function of concentration does not lead to analytical expressions in closed form, though the possibility of computer solutions exist. Analytical solutions, therefore, have proceeded by assuming the Langmuir isotherm. In this case,

$$\bar{c}(0, s) = L \left\{ c^0 - \frac{\theta(t)}{k[1 - \theta(t)]} \right\} \tag{32}$$

and equating the two expressions [equations (31) and (32)] for $\bar{c}(0, s)$, one has

$$(s/D)^{1/2}\bar{\Gamma}(s) = L \left\{ c^0 - \frac{\theta(t)}{k[1 - \theta(t)]} \right\}$$

$$= \int_0^\infty e^{-st} \left\{ c^0 - \frac{\theta(t)}{k[1 - \theta(t)]} \right\} dt \tag{33}$$

This integral equation has been solved [9], but the solutions involve rather cumbersome recurrence relations. This introductory treatment will concern itself, therefore, with two important special cases.

Special Case I

$$c^0 \gg \frac{\theta(t)}{k[1 - \theta(t)]} \tag{34}$$

The general Langmuir-based equation (33) reduces under this condition to

$$(s/D)^{1/2} \bar{\Gamma}(s) = L(c^0) = c^0/s$$

or

$$\bar{\Gamma}(s) = D^{1/2}c^0 s^{-3/2} \tag{35}$$

What is required, however, is the variation of the surface concentration Γ with time. That is, one seeks $\Gamma(t)$, rather than $\bar{\Gamma}(s)$. The recovery of $\Gamma(t)$ from $\bar{\Gamma}(s)$ involves a process of Laplace transformation in reverse, or a so-called inverse transformation. Tables [10] of transforms and inverse transforms are available, and by reference to them one finds that

$$\Gamma(t) = (2D^{1/2}\pi^{-1/2}c^0) t^{1/2} \tag{36}$$

The question is: To what physical situations does this special case, viz., $c^0 \gg \theta(t)/k[1 - \theta(t)]$ correspond? Let us rewrite the special case in the form $c^0 \gg c'(0, t)$ by making use of the Langmuir isotherm—equation (10). One has to analyze, therefore, what physical situations correspond to a negligible value of $c'(0, t)$, i.e., the concentration of adsorbate at the $x = 0$-plane. By stipulating that the adsorbed species does not undergo charge-transfer reactions, the possibility of external control on the diffusion process has been eliminated. The controlling of the mass transport is achieved by the adsorption isotherm. When the system has attained a time-independent state (adsorption equilibrium), diffusion must stop. This occurs when $c'(0, t) = c^0$. Hence, the only condition under which $c'(0, t) \ll c^0$ is valid is *for times short in comparison with the time to reach equilibrium.*

It is possible to cast the short-time special case into other convenient forms. Thus, by using the linear isotherm* $\Gamma(t) = Kc'(0, t)$, where $k = K\Gamma_{max}$, one obtains for the equilibrium situation

$$\Gamma(t \to \infty) = Kc'(0, t \to \infty) \tag{37}$$

or

$$\Gamma_{eq} = Kc'_{eq} = Kc^0 \tag{38}$$

Substituting for c^0 in equation (36), it is found that

$$\frac{\Gamma(t)}{\Gamma_{eq}} = \frac{2D^{1/2}}{K\pi^{1/2}} t^{1/2} \tag{39}$$

If the physical situations correspond to the special case discussed here, the $\Gamma(t)$ vs. $t^{1/2}$ plot will be linear for short times.

It is also possible to use equation (39) to define a time τ required to attain the condition $\Gamma(t)/\Gamma_{eq} = 1$. The result is

$$\tau^{1/2} = \frac{K\pi^{1/2}}{2D^{1/2}} \tag{40}$$

or

$$\tau = \frac{K^2\pi}{4D} \tag{41}$$

It must be noted, however, that since the $\Gamma(t)$ vs. $t^{1/2}$ law is valid only for short times, there will be a deviation for long times, i.e., when $\Gamma(t) \to \Gamma_{eq}$, and, therefore, τ is an extrapolated value.

* The linear isotherm is obtained from equations (10) and (9) by assuming $\theta \to 0$, i.e., $1 - \theta \approx 1$, in which case $\theta(t) \approx kc'(0, t) = \Gamma(t)/\Gamma_{max}$.

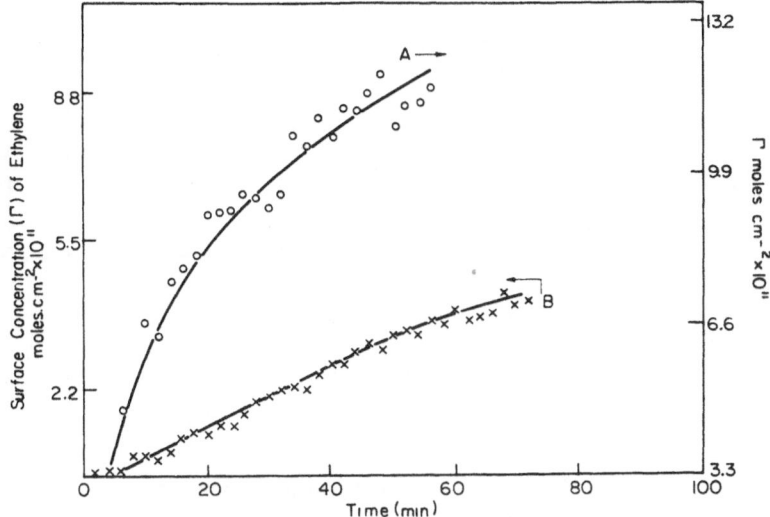

Fig. 3. Variation in surface concentration (measured in counts per minute from "tagged" ethylene) as a function of time. Curve *A*—ethylene bulk concentration of $(2.4–4.5) \times 10^{-6} M$. Curve *B*—ethylene bulk concentration of $(0–5.8) \times 10^{-7} M$.

Excellent confirmation of this analysis has come from the radio-tracer study [11] of the electrosorption of ethylene from $1\ N H_2 SO_4$ on Pt-plated gold electrodes. The variation of the surface concentration with time is shown in Fig. 3, but the agreement with the theory— equation (36)—is better seen by plotting Γ vs. $t^{1/2}$, in which case a straight-line plot is obtained (Fig. 4).

The slope of the straight-line plot can be used—cf. equation (39)— to get at the equilibrium constant, $k = K/\Gamma_{max}$, for the adsorption reaction. When this equilibrium constant is compared with the value obtained from steady-state measurements, it is seen that there is fair agreement—a further check on the theory (Table I). Further support for the $\Gamma(t)$ vs. $t^{1/2}$ law can be found in studies on the adsorption of benzene [12] and propane [13] on platinum electrodes.

It may appear that the times (of the order of minutes) obtaining in the above-mentioned experiments are far too long to avoid natural upsetting of a simple diffusion treatment of the adsorption problem. It has been shown [14], however, that even under conditions of stirring or forced convection the response of the system to a unit step is exactly the same as in the pure diffusion case *up to a critical time* t_c. Thereafter,

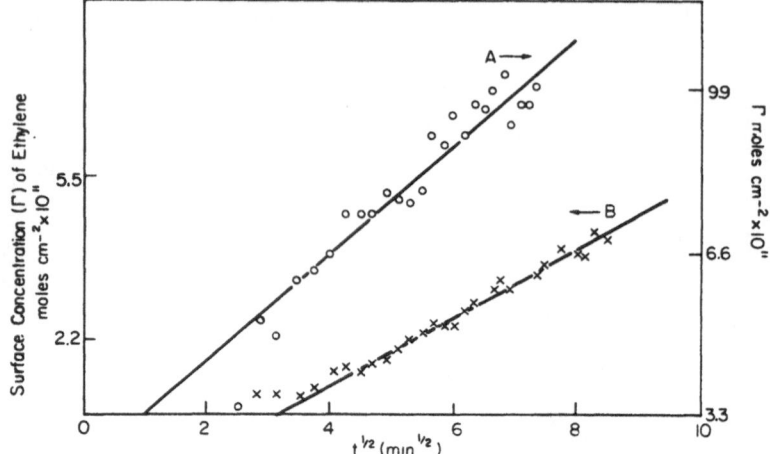

Fig. 4. Same as Fig. 3, but plotted vs. (time)$^{1/2}$.

Table I

Temperature °C	$10^8 \times k$ cm^3/mole			
	From equation (41)	From equation (39)	Average	From independent steady-state measurement
30	9.7	9.5	9.6	7.4
70	10.0	5.8	7.9	9.2

forced convection or the onset of natural convection alters the response. Hence, upto a time t_c , the diffusion analysis and its final expression for $\Gamma(t)$ can be used. The question is: What is the value of t_c ? The value can emerge only from a detailed analysis of the hydrodynamic problem with all its attendant difficulties. In the absence of such an analysis, one would normally think of extrapolating from the dropping mercury electrode situation, where it is known that natural convection sets in after 10 to 15 sec. But, in the case of diffusion-controlled adsorption onto the solid electrodes of the above experiments, there are several distinctive features. First, a horizontal electrode is used with the consequent minimization of natural convection. Second, the maximum amount of material that can be removed from the system by adsorption is only that corresponding to a monolayer. The density differences resulting from the removal of this small amount of material are completely negligible compared to those arising from the continuous removal by electrolysis at the dropping mercury electrode. Hence, one cannot conclude unambiguously that ideas of t_c gleaned at the dropping mercury electrode are extendable in any straightforward manner to the case of adsorption of electroinactive neutral molecules on solid electrodes.

Special Case II

$$\theta \ll 1 \quad \text{or} \quad \frac{\theta}{k(1 - \theta)} \approx \frac{\theta}{k} \quad \text{or} \quad c^0 - \frac{\theta}{k(1 - \theta)} \approx c^0 - \frac{\theta}{k} \quad (42)$$

In this case,

$$L \left\{ c^0 - \frac{\theta(t)}{k[1 - \theta(t)]} \right\} \approx L[c^0 - \theta(t)/k]$$

$$\approx L \left\{ c^0 - \frac{1}{k\Gamma_{max}} \Gamma(t) \right\}$$

$$\approx \frac{c^0}{s} - \frac{1}{K} \bar{\Gamma}(s) \quad (43)$$

where, as before, K is written for $k\Gamma_{max}$. Inserting this expression into equation (33), one obtains

$$(s/D)^{1/2} \bar{\Gamma}(s) = \frac{c^0}{s} - \frac{1}{K} \bar{\Gamma}(s)$$

i.e.,

$$\bar{\Gamma}(s) \left[(s/D)^{1/2} + \frac{1}{K} \right] = \frac{c^0}{s}$$

or

$$\bar{\Gamma}(s) D^{-1/2} \left[s^{1/2} + \frac{D^{1/2}}{K} \right] = \frac{c^0}{s}$$

which leads to

$$\bar{\Gamma}(s) = c^0 D^{1/2} \frac{1}{s(b + s^{1/2})}$$

$$= c^0 K \frac{b}{s(b + s^{1/2})} \quad (44)$$

where

$$b = \frac{D^{1/2}}{K}$$

The inverse transformation is done with the aid of the relation

$$\alpha \left[1 - e^{b^2 t}\, \text{erfc}(bt^{1/2})\right] = \frac{b}{s(b + s^{1/2})} \tag{45}$$

and one gets

$$\Gamma(t) = c^0 K \left[1 - \exp(Dt/K^2)\, \text{erfc}\left(\frac{D^{1/2} t^{1/2}}{K}\right)\right] \tag{46}$$

One can resort to an approximation at this stage. All terms after the first can be dropped in the expansion

$$\exp x^2\, \text{erfc}\, x = \frac{1}{x/\pi}\left[1 - \frac{1}{2x^2} + \frac{1.3}{(2x^2)^2} \cdots\right] \tag{47}$$

where $x = (D^{1/2} t^{1/2})/K$. This approximation is valid to within 1 % for $t > 50K^2/D$. One obtains

$$\Gamma(t) = c^0 K \left[1 - \frac{K}{(Dt\pi)^{1/2}}\right] \tag{48}$$

It has been repeatedly stated that in the diffusion-controlled adsorption of electroinactive molecules the diffusion process is programmed by the adsorption isotherm. This point is brought out clearly by equation (48), which shows that $\Gamma(t)$ depends on the constant $K = k\Gamma_{max}$, which is a characteristic of the adsorption process.

Since the special case under consideration corresponds to the use of the linear isotherm, i.e., $\theta \ll 1$, and hence $\theta = kc = \Gamma(t)/\Gamma_{max}$, the factor $c^0 K$ can be replaced by $\Gamma_{eq} = \Gamma(t \to \infty)$. Thus,

$$\Gamma(t) = \Gamma_{eq}\left[1 - \frac{K}{(Dt\pi)^{1/2}}\right] \tag{49}$$

As in Special Case I, viz., $c^0 \gg c'(0)$, one has to ask what physical conditions correspond to Special Case II, viz.,

$$\frac{\theta(t)}{k[1 - \theta(t)]} \approx \frac{\theta(t)}{k}$$

The physical condition is that if $\theta_{eq} = \theta(t \to \infty) \ll 1$, then it will certainly be true that $0 < \theta(t) < \theta_{eq} \ll 1$. This is the condition of the linear isotherm.

The experimental testing of equation (49) can be done by recasting it into the following form:

$$t^{1/2} = \frac{K}{(\pi D)^{1/2}} \left[1 - \frac{\Gamma(t)}{\Gamma_{eq}} \right]^{-1} \tag{50}$$

or

$$t = \frac{K^2}{\pi D} \left[1 - \frac{\Gamma(t)}{\Gamma_{eq}} \right]^{-2} \tag{51}$$

and then defining a rise time τ corresponding to the condition

$$\Gamma(t) = 0.97 \, \Gamma_{eq} \tag{52}$$

In other words,

$$\tau = \frac{K^2}{\pi D} 0.03 \tag{53}$$

The calculated and measured rise times for the adsorption of dibutyl ketone and some other substances on mercury electrodes is shown [15] in Table II. The agreement appears to be fair.

Table II

	Rise time, τ (in sec)	
Substance	Calculated	Observed
$(C_4H_9)_2CO$	220	200–400
$C_6H_5 \cdot SH$	200	100–300
$\alpha\text{-}C_{10}H_7 \cdot NH_3^+$	45	60

5 MASS TRANSPORT-CONTROLLED ELECTROSORPTION UNDER STIRRED CONDITIONS

The treatment that has been presented so far has restricted itself to situations in which the adsorbate moves up to the electrode by pure diffusion. However, when the transport occurs by a process of forced convection, then a simple analysis can be made in terms of the concept of the Nernst diffusion layer. According to this concept, the concentration variation with distance away from the electrode can be linearized (Fig. 5), and it can be assumed that, effectively speaking, the concentration varies linearly from the value at $x = 0$ to the bulk value over a distance δ, the diffusion layer thickness. Then, the steady-state diffusion flux J is related to the concentration gradient $[c^0 - c(0, t)]/\delta$ through Fick's first law:

$$J = D \frac{c^0 - c'(0, t)}{\delta} \qquad (54)$$

Further, the condition of flux continuity can be used to obtain the expression

$$\frac{d\Gamma(t)}{dt} = J = D \frac{c^0 - c'(0, t)}{\delta} \qquad (55)$$

The use of Fick's first law for non-steady-state conditions is tantamount to considering that steady state prevails over infinitesmal time intervals dt. From the linear isotherm [equation (9)],

$$\theta = kc' = \frac{\Gamma}{\Gamma_{\max}}$$

or

$$\Gamma = kc'$$

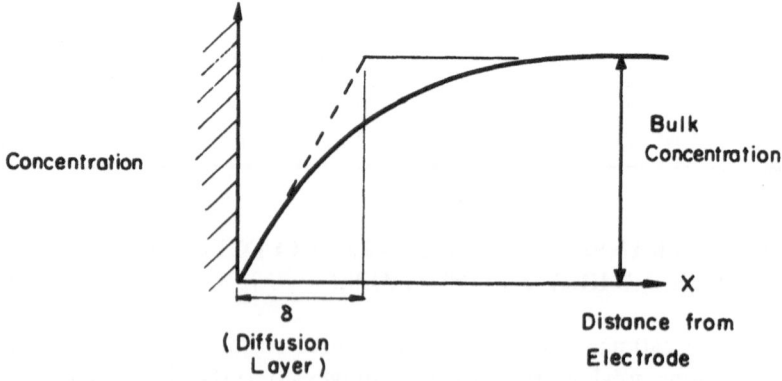

Fig. 5. Concentration profile to illustrate concept of diffusion layer.

Using this relation, one gets

$$c^0 - c' = (1/K)[\Gamma_{max} - \Gamma(t)] \tag{56}$$

and, therefore, from equation (54) one has

$$\frac{d\Gamma(t)}{dt} = \frac{D}{K\delta}[\Gamma_{max} - \Gamma(t)] \tag{57}$$

Integration of this equation leads to

$$\ln[\Gamma_{max} - \Gamma(t)] = \frac{Dt}{K\delta} + B \tag{58}$$

where B is the integration constant that is evaluated from the condition that at $t = 0$, $\Gamma(t) = \Gamma(0) = 0$. It is found that

$$B = \ln \Gamma_{max} \tag{59}$$

Hence,

$$\ln\left[1 - \frac{\Gamma(t)}{\Gamma_{max}}\right] = \frac{-Dt}{K\delta}$$

or

$$\Gamma(t) = \Gamma_{max}\left[1 - \exp\left(-\frac{Dt}{K\delta}\right)\right] \tag{60}$$

This result contains an important approximation in that the diffusion layer thickness δ is considered a constant during the integration of the differential equation, despite the fact that δ must be a variable before the attainment of the true steady state.

Equation (60) has been used [16] to account for the time effects observed in anodic oxidation of hydrocarbons. The argument is as follows. Dividing both sides of equation (60) by Γ_{max}, one gets

$$\theta(t) = \theta_{max}\left[1 - \exp\left(-\frac{Dt}{K\delta}\right)\right] \tag{61}$$

The familiar expression for the oxidation current density is

$$i = k'(1 - \theta) \exp\left(\frac{\alpha FV}{RT}\right) \tag{62}$$

Combining equations (61) and (62), one has

$$i(t) = k'\left\{1 - \theta_{max}\left[1 - \exp\left(-\frac{Dt}{K\delta}\right)\right] \exp\left(\frac{\alpha FV}{RT}\right)\right\} \tag{63}$$

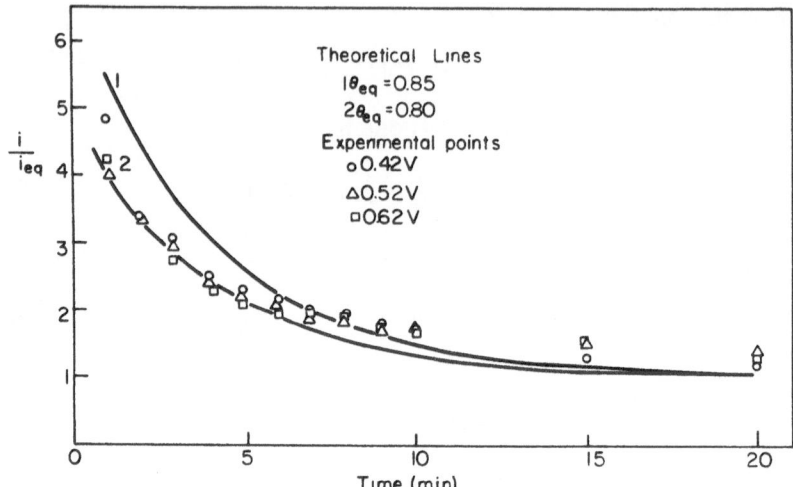

Fig. 6. Long-time effect in the case of ethylene oxidation interpreted on the basis of equation (64).

If the current density at $t \to \infty$ is denoted by $i(t \to \infty)$, then,

$$\frac{i(t)}{i(t \to \infty)} = \frac{1 - \theta_{max}\left[1 - \exp\left(-\dfrac{Dt}{K\delta}\right)\right]}{1 - \theta_{max}} \qquad (64)$$

Using the experimental values of θ_{max} and $D/K\delta$, one can predict the variation of $i(t)/i(t \to \infty)$ with time. The comparison between theory and experiment is shown in Fig. 6.

REFERENCES

1. Gileadi and Conway, *Modern Aspects of Electrochemistry*, Vol. III, Chap. V, Bockris and Conway, eds., Butterworth, Washington, D. C. (1964).
2. Delahay, *Double Layer and Electrode Kinetics*, John Wiley & Sons, New York (1965).
3. Bockris and Reddy, *A Course in Modern Electrochemistry*, Vol. II, Plenum Press, New York, in press.
4. Conway, Bockris, and Lovrecek, *Compt. Rend.* CITCE V1, 207 (1955).
5. Fischer, *Elektrolytische Abscheidung and Elektrokristallisation von Metallen*, Springer, (1954).

6. Delahay and Trachtenberg, *J. Am. Chem. Soc.* **79**, 2355 (1957).
7. Delahay and Fike, *J. Am. Chem. Soc.* **80**, 2628 (1958).
8. Hansen, *J. Phys. Chem.* **64**, 637 (1960).
9. Reinmuth, *Anal. Chem.* **65**, 473 (1961).
10. *Handbook of Chemistry and Physics*, Chemical Rubber Publishing Company.
11. Gileadi, Rubin, and Bockris, *J. Phys. Chem.* **69**, 3335 (1965).
12. Heiland, Gileadi, and Bockris, *J. Phys. Chem.* **70**, 1207 (1966).
13. Brummer, Ford, and Turner, *J. Phys. Chem.* **69**, 3424 (1965).
14. Nanis, *Ann. N.Y. Acad. Sci.* **105**, 667 (1963).
15. Blomgren, Bockris, and Jesch, *J. Phys. Chem.* **65**, 2000 (1961).
16. Bockris, Wroblowa, Gileadi, and Piersma, *Trans. Faraday Soc.* **61**, 2531 (1965).

Chapter 4

Oxygen Adsorption and Oxide Formation on Electrodes

M. A. Genshaw

1. INTRODUCTION

The purpose of this discussion is to show the importance of oxygen on electrode surfaces and to try to give some insight into the structure and mechanism of formation of these species.

The results of work in both gas phase and solution phase will be considered in this discussion, as measurements in the gas phase can supplement the limited information obtainable in measurements made in solution.

In the electrochemical literature as a whole, little distinction has been made between adsorbed oxygen and oxide. A surface film is discussed as "chemisorbed oxygen" when it may be oxide. For this discussion to avoid confusion, we will define oxide to include any species in which the metal atoms have moved from their lattice positions to form the species. Thus, chemisorbed oxygen will only include formations of bonds between the substrate and oxygen without movement of the substrate atoms.

2. THERMODYNAMICS OF OXIDATION

From a thermodynamic viewpoint, oxide should readily form on most metals. This is apparent from the free energies of formation of oxides, as gold is the only metal which does not have an oxide formed with a negative free energy of formation. Thus, all other metals should spontaneously oxidize in the presence of air. However, it is equally

73

apparent that all metals do not oxidize rapidly, as many metals are used in products that are durable.

Oxygen may react with surfaces in two ways. In the first, physical adsorption, no chemical bonds are formed between the oxygen and the substrate. In the second, chemical bonds are formed between the oxygen and the substrate. This leads to chemisorption or to compound formation.

3. CHARACTERISTICS OF PHYSICALLY ADSORBED OXYGEN

Physical adsorption is a weak adsorption. The heat of adsorption amounts to no more than a few kilocalories per mole. The bonding is by van der Waals forces. The adsorption is molecular. It occurs near the boiling point of oxygen ($-183°C$). The adsoprtion has no activation energy, and several layers of physically adsorbed oxygen may form on the surface.

4. CHARACTERISTICS OF CHEMISORBED OXYGEN

Chemisorption is a strong adsorption. Dissociation of the oxygen occurs and ionic or covalent bonds are formed. The heat of adsorption is at least 15 kcal/mole. Chemisorption may occur at any temperature, although a heat of activation for adsorption may lead to no apparent adsorption at low temperatures—this is a kinetic limitation of adsorption, not a thermodynamic limitation. Chemisorption is usually irreversible at room temperature. Usually only a monolayer or less is adsorbed, although chemisorption may be the first step in a process leading to multilayer formation.

5. THE OCCURRENCE OF PHYSICAL ADSORPTION

One would expect physical adsorption at low temperatures on metals. However, even at the boiling point of oxygen, oxygen is chemisorbed on metals [1]. Physical adsorption will occur on the partially oxidized metal after the chemisorption ceases. Physical adsorption occurs on inert substrates, such as carbon, at liquid air temperature.

6. THE OCCURRENCE OF CHEMISORPTION

Chemisorption, even when restricted to the definition used here, has been experimentally observed. To isolate the chemisorption from oxide formation, it is necessary to work at very low partial pressures of oxygen. The diffusion of oxygen to the surface then becomes slow, so that the kinetics of chemisorption are determined by the diffusion rather than by a slow adsorption step.

The structure of the chemisorbed layer can be determined by low-energy electron diffraction. The structures observed conform to the metal lattice. However, the structures are not always simple structures corresponding to the underlying lattice, but may be rotated with respect to the metal lattice.

The coverage of the metal with chemisorbed oxygen is apparently not determined solely by geometric factors, as a maximum coverage with chemisorbed oxygen may be reached. On the 111 face of platinum the structure with the highest coverage observed corresponds to 1/2 coverage [2]. On rhodium, the chemisorbed oxygen assumes a close-packed structure that corresponds to a 7/8 coverage (oxygen atoms to surface rhodium atoms) [3]. On nickel, a structure corresponding to Ni_3O forms. When more oxygen is adsorbed, the oxide NiO is formed [4].

It is probable that the chemisorbed oxygen forms bonds that are partially ionic. This is seen in the large changes in work function of metals upon oxygen adsorption [1,4]. It is also directly shown in the observed changes of heats of adsorption with coverage [5].

Chemisorption is the first stage of oxidation of most metals. To understand the subsequent steps, we must consider the mechanism of oxide growth.

7. MECHANISM OF OXIDE GROWTH

The mechanism of growth will be determined to a considerable extent by the electronic properties of the oxide. Three types of oxides exist: (1) nonconducting oxides, (2) semiconducting oxides, and (3) conducting oxides.

If a current is passed through the oxide, an electric field is built up across the oxide. In the case of nonconducting oxides, this field increases until ionic conduction or dielectric breakdown occurs. For conducting oxides, the field across the oxide must be quite small. For semiconducting oxides, the field is between the values for non-conductors and conductors. We will consider mechanisms that have been proposed for each type of oxide.

7.1. Growth of Nonconducting Oxides

With nonconducting oxides it is evident that a large electric field may arise across the oxide. This may result either from an externally applied field or from the adsorption of oxygen on the oxide surface. For the former case, the potential difference may be any value; for the latter, it may not exceed a few volts, as it may not exceed the difference in electronegativities. This field will promote ionic conduction in the direction of the field. Let us consider the barrier restraining movement of an ion. An ion will normally be near the bottom of the potential well formed by interaction with neighboring ions (Fig. 1). For movement, the ion must pass over this free energy barrier. The rate of passage of ions over this barrier is

$$\frac{dL}{dt} = A \exp\left(-\frac{\Delta G_0^{\ddagger}}{RT}\right) \tag{1}$$

where dL/dt is the rate of oxide growth; A is a constant; $\Delta G_0^{\ddagger}$ is the standard free energy of activation for the oxide growth; R the gas

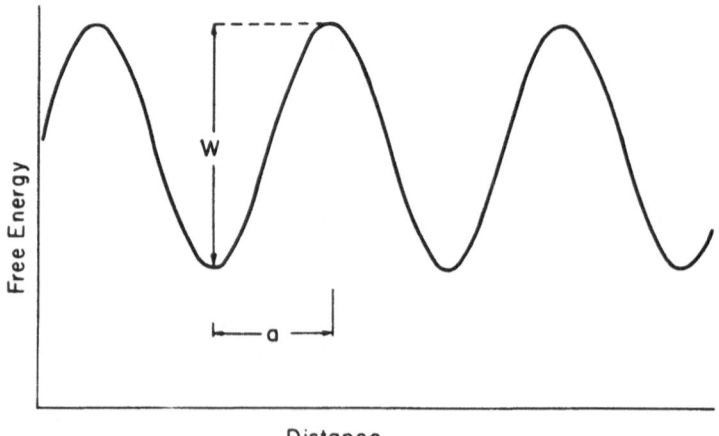

Fig. 1. Free energy of an ion as a function of distance.

constant; and T is the absolute temperature. The electric field acts to accelerate the passage of ions in the direction of the field. We will now consider the net flux into the film. This flux is the difference between the flux in the forward and reverse directions and may be expressed in terms of the rate of growth of the oxide as the growth is proportional to the number of ions entering the oxide layer:

$$\left(\frac{dL}{dt}\right)_{net} = A \exp\left(-\frac{\Delta G_0^{\ddagger}}{RT} + \frac{naFE}{RT}\right)$$

$$- A \exp\left(-\frac{\Delta G_0^{\ddagger}}{RT} - \frac{naFE}{RT}\right) \tag{2}$$

$$\left(\frac{dL}{dt}\right)_{net} = \frac{A}{2} \exp\left(-\frac{\Delta G_0^{\ddagger}}{RT}\right) \sinh\left(\frac{naFE}{RT}\right) \tag{3}$$

Here n is the relative charge on the moving ion; a is the distance from the minimum to the peak of the free energy barrier (Fig. 1); and E is the electric field:

$$E = V/L \tag{4}$$

Here V is the total potential across the oxide and L is the thickness of the oxide.

(a) Case 1

$$\frac{naFE}{RT} \ll 1 \tag{5}$$

$$\frac{dL}{dt} = \frac{2A\,naFE}{RT} \exp\left(-\frac{\Delta G_0^{\ddagger}}{RT}\right) \tag{6}$$

Substituting equation (4) into (6), we obtain

$$\frac{dL}{dt} = \frac{2A\,naFV}{LRT} \exp\left(-\frac{\Delta G_0^{\ddagger}}{RT}\right) \tag{7}$$

Integrating this equation, we have

$$L^2 = \frac{4A\,naFVt}{RT} \exp\left(-\frac{\Delta G_0^{\ddagger}}{RT}\right) + \text{const} \tag{8}$$

If we define the time as zero when the thickness is zero, the constant of integration is zero. Then we see that

$$L \propto t^{1/2} \tag{9}$$

This equation is the parabolic growth law that is frequently observed. However, as we shall see later, it also arises when simple diffusion is considered.

(b) *Case* 2

$$\frac{naFE}{RT} \gg 1 \qquad (10)$$

From equation (2):

$$\frac{dL}{dt} = A \exp\left(\frac{\Delta G_0^{\ddagger}}{RT}\right) \exp\left(\frac{naFE}{RT}\right) \qquad (11)$$

$$\exp\left(-\frac{naFE}{RT}\right) dL = A \exp\left(\frac{\Delta G_0^{\ddagger}}{RT}\right) dt \qquad (12)$$

Substituting equation (4) into (12), we obtain:

$$\exp\left(-\frac{naFV}{LRT}\right) dL = A \exp\left(\frac{\Delta G_0^{\ddagger}}{RT}\right) dt \qquad (13)$$

Cabrera and Mott [7] have solved this equation for the condition

$$L \ll \frac{naFV}{RT} \qquad (14)$$

They give the approximate solution

$$\frac{L^2 RT}{naFV} \exp\left(-\frac{naFV}{RTL}\right) = A \exp\left(\frac{\Delta G_0^{\ddagger}}{RT}\right) t + \text{const} \qquad (15)$$

Defining the thickness as zero when the time is zero, the constant of integration is zero. Taking the logarithm of each side, we obtain

$$2 \ln L + \ln \frac{RT}{naFV} - \frac{naFV}{RTL} = \ln A + \frac{\Delta G_0^{\ddagger}}{RT} + \ln t \qquad (16)$$

If we neglect the change in $2 \ln L$ in comparison with the change in $naFV/RTL$, the equation becomes

$$\frac{naFV}{RTL} = \ln \frac{RT}{naFV} - \ln A - \frac{\Delta G_0^{\ddagger}}{RT} - \ln t \qquad (17)$$

The growth follows an inverse logarithmic law.

To obtain the relation between thickness and potential, we may neglect the change in $\ln 1/V$ in comparison with the change in $naFV/RTL$. Equation (17) becomes

$$\frac{naFV}{RTL} = \ln \frac{RT}{naF} - \ln A - \frac{\Delta G_0^{\ddagger}}{RT} - \ln t \qquad (18)$$

From (18) we see that, at constant time, L is linearly dependent on V. For actual measurements, when thickness is measured after a period of time has passed, say 1 hr the rate of change in thickness per minute will become quite small when compared with the total thickness, as the dependence on time is logarithmic.

The linear dependence of thickness on potential has frequently been observed, particularly with the valve metals.

7.2. Growth of Semiconducting Oxides

Uhlig [8] has derived kinetics for growth of semiconducting oxides. He assumes that the oxide has a different work function than the metal. The oxide will then acquire a charge. If there are relatively few sites in the oxide for the excess charge, the excess charge must be spread out over a considerable region. This is in contrast to the case of metals in contact, where the electric charge is confined to a region of atomic dimensions.

For a first approximation, we shall assume that the charge distribution is uniform. Over a distance L_1, the total change must compensate the opposite charge on the metal.

From Poisson's equation, we have

$$\frac{d^2\Psi}{dL^2} = \frac{4\pi\rho e}{\epsilon} \qquad (19)$$

where Ψ is the potential with respect to the bulk of the oxide; ρ is the density of trapped electrons (or holes) in the oxide; e is the electronic charge; and ϵ is the dielectric constant. Ψ is the change in potential at the metal-oxide interface, due to the space charge in the oxide.

Integrating (19), we obtain

$$\frac{d\Psi}{dL} = \frac{4\pi\rho e}{\epsilon} L + C_1 \qquad (20)$$

If we consider that the total space charge is confined to a region of thickness L_1, we may define the boundary condition $dV/dL = 0$, when $L = L_1$. Hence,

$$C_1 = \frac{-4\pi\rho e L_1}{\epsilon} \tag{21}$$

Integrating again,

$$\Psi = \frac{-4\pi\rho e}{\epsilon}\left(L_1 L - \frac{L^2}{2}\right) + C_2 \tag{22}$$

Since we have defined Ψ as the potential charge due to the space charge in the oxide, when $L = 0$, $\Psi = 0$. Hence, the constant C_2 is zero.

We will assume that the rate of the oxide growth is controlled by the rate of electron transfer. The electron current across the contact of a metal and a semiconducting oxide has been found to follow an equation of the form [9]

$$i = B \exp\frac{-e\varphi}{kT}\left(\exp\frac{e\chi}{kT} - 1\right) \tag{23}$$

where B is a constant; φ is the work function; and χ is the applied voltage across the metal–oxide interface.

If we neglect the small portion of the total charge transferred that is involved in the space charge, the rate of growth of the oxide is proportional to the rate of electron transfer:

$$C\frac{dL}{dt} = i = B\exp\left(\frac{-e\varphi}{kT}\right)\left(\exp\frac{e\chi}{kT} - 1\right) \tag{24}$$

Since we will be considering potentials where $\exp(e\chi/kT) \gg 1$, we may neglect the 1 in the term $[\exp(e\chi/kT) - 1]$. The potential drop across the metal–oxide interface will be proportional to the potential set up by the space charge. Thus, we have

$$\chi = \chi^0 + \Psi \tag{25}$$

where χ^0 is the potential drop that would hypothetically exist if the oxide thickness were zero. Substituting (25) and (22) into (24), we obtain

$$\frac{CdL}{dt} = B\exp\left(\frac{-e\varphi}{kT}\right)\exp\frac{e\chi^0}{kT}\exp\frac{-4\pi\rho e^2}{\epsilon kT}\left(L_1 L - \frac{L^2}{2}\right) \tag{26}$$

Defining a new constant:

$$g = \frac{B}{C}\exp\left(\frac{-e\varphi}{kT}\right)\exp\left(\frac{e\chi^0}{kT}\right)\exp\left(\frac{eC_1}{kT}\right) \tag{27}$$

we then have

$$\frac{dL}{dt} = g \exp\left[\frac{-4\pi\rho e^2}{\epsilon kT}\left(L_1 L - \frac{L^2}{2}\right)\right] \tag{28}$$

If we now consider films where L is much less than L_1, we may neglect the term in L^2. Then,

$$\frac{dL}{dt} = g \exp\frac{-4\pi\rho e^2}{\epsilon kT} L_1 L \tag{29}$$

Integrating, we obtain

$$\exp\frac{4\pi\rho e^2}{\epsilon kT} L_1 L = gt + \text{const} \tag{30}$$

The constant of integration is equal to 1, since the thickness is zero when the time is zero. The dependence of L on time is then

$$L = \frac{\epsilon kT}{4\pi\rho e^2 L_1} \ln\left(gt + 1\right) \tag{31}$$

When we have

$$Dt \gg 1 \tag{32}$$

the growth follows the logarithmic law:

$$L = \frac{\epsilon kT}{4\pi\rho e^2 L_1} \ln gt \tag{33}$$

7.3. Growth of Conducting Oxides

For oxides that are conducting, a mechanism different from the high-field mechanisms is required. Due to the electronic conduction, the electric field across a conducting oxide must be very small. Its effect on the oxide growth must be correspondingly small.

In the mechanisms we have considered thus far, the electric field has played a dominent role in the ion migration. In the absence of this field, we might expect that the ion migration step is still the slowest. We would then have the rate expression

$$\frac{dL}{dt} = A \exp\left(-\frac{\Delta G_0^{\ddagger}}{RT}\right) \tag{34}$$

The growth would then be at a constant rate.

Fig. 2. Illustration of place exchange.

An experimental factor that should be considered is the observation that the oxide thickness at room temperature is often less than 100 Å [7]. In the thickness range of a few layers, we might expect that the energy of activation for ion movement is dependent on the thickness.

A mechanism that has been proposed for the growth of thin oxide films is the place exchange mechanism [10]. The model proposed for place exchange is shown in Fig. 2.

In Fig. 2, we illustrate the steps following oxygen adsorption. The oxygen is first chemisorbed on the surface. As discussed earlier, the chemisorbed oxygen is probably somewhat ionic. Due to the attraction of the oxygen toward the metal and the repulsion of neighboring groups, there is a stimulus for the metal atom to exchange places with the oxygen atom.

After the exchange has occurred, the final arrangement is more stable than the original configuration. Thus, there is little tendency for the reverse reaction to occur. Also, the neighboring groups have also been stabilized because the repulsive interaction has been replaced with an attractive interaction.

Further oxygen can chemisorb on the metal atoms exposed by place exchange.

In Fig. 3, a second property of the place exchange is illustrated. Once the oxide thickness is greater than a monolayer, in order to exchange groups, a series of simultaneous place exchanges occur. This also leads to an increase in the energy of activation with thickness.

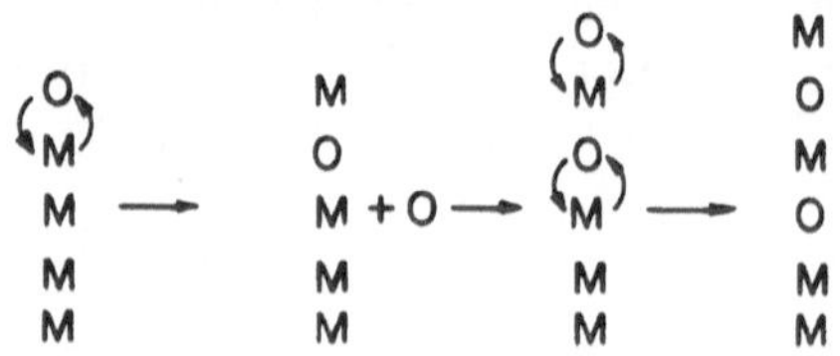

Fig. 3. Illustration of simultaneous place exchange.

We shall assume that the free energy of activation increases linearly with thickness:

$$\Delta G^{\ddagger} = \Delta G_0^{\ddagger} + bL \tag{35}$$

Then the rate expression becomes

$$\frac{dL}{dt} = A \exp\left(-\frac{\Delta G_0^{\ddagger} + bL}{RT}\right) \tag{36}$$

$$\exp\left(\frac{bL}{RT}\right) dL = A \exp\left(-\frac{\Delta G_0^{\ddagger}}{RT}\right) dt \tag{37}$$

Integrating and using the boundary condition of $L = 0$ at $t = 0$, we have

$$\exp\frac{bL}{RT} = \frac{RTA}{b} \exp\left(-\frac{\Delta G_0^{\ddagger}}{RT}\right) t + 1 \tag{38}$$

If we consider times great enough so that

$$\frac{RTA}{b} \exp\left(-\frac{\Delta G_0^{\ddagger}}{RT}\right) t \gg 1 \tag{39}$$

we may neglect the 1 in (38). Upon taking logarithms, equation (39) becomes

$$L = \frac{RT}{b} \ln t + \frac{RT}{b} \ln \frac{RTA}{b} - \frac{\Delta G_0^{\ddagger}}{b} \tag{40}$$

Again, we have a logarithmic growth law.

In the case of oxide growth in solution, we also find a potential dependence. Here we may consider that the free energy of the chemisorbed oxygen is potential dependent. This potential dependence arises because the chemisorbed oxygen is produced by an equilibrium discharge process. Hence,

$$\Delta G^{\ddagger} = \Delta G_0^{\ddagger} - cFE \tag{41}$$

where c is a constant and E is the electrode potential. Since we have not defined the potential scale used, the value of A will be dependent on the potential scale. Substituting (41) into (40), we obtain

$$L = \frac{RT}{b} \ln t + \frac{RT}{b} \ln \frac{RTA}{b} - \frac{\Delta G_0^{\ddagger}}{b} + \frac{cFE}{b} \tag{42}$$

Here, we see that at constant potential the growth again follows the logarithmic law, as has been experimentally observed [11,12]. At constant time, the thickness increases linearly with potential. The linear thickness–potential relation has been observed for a number of metals [13].

7.4. Growth of Oxides by Diffusion

The mechanisms of oxide formation that result in logarithmic growth laws provide a path for very rapid initial growth of oxides. However, the rate of growth decreases rapidly and in many cases becomes very slow, at times on the order of minutes. Then, another mechanisms of oxide formation may become faster. We will now consider growth due to simple diffusion.

There is a concentration gradient across the oxide. In the previous derivations we have neglected this because it is a small driving force in comparison with the other forces considered.

Now we also must consider the back reaction, which was neglected in the previous derivations.

If we now consider that the growth occurs by simple diffusion, we may apply Fick's first law:

$$\frac{dC}{dt} = D \frac{dC}{dL} \tag{43}$$

where C is the concentration of metal ions in the oxide and D is the diffusion coefficient for metal ions in the oxide. Let us consider that the concentration gradient is linear. Then

$$\frac{dC}{dt} = D \left(\frac{C_i - C_0}{L} \right) = H \frac{dL}{dt} \tag{44}$$

Here, C_i and C_0 are the concentration of metal ions in the oxide at the metal–oxide and oxide–gas (or oxide–solution) interfaces, respectively, and H is the proportionality constant relating concentration to thickness. The diffusion flux is proportional to the rate of growth of the oxide, as all of the metal ions and oxygen ions which diffuse into the oxide remain in it:

$$\frac{dC}{dt} = H \frac{dL}{dt} = \frac{D(C_i - C_0)}{L} \tag{48}$$

Integrating,

$$L^2 = \frac{2D}{H} (C_i - C_0)t \qquad (49)$$

This equation is the well-known parabolic growth law. Parabolic growth has been observed for oxide formation on metals that normally follow the logarithmic growth, when the kinetics have been determined in the very slow growth region (at room temperature, measurements begin after 2 hr of growth) [14]. The parabolic law is more frequently observed for growth at elevated temperatures when the ions have greater mobility [7].

8. WHY IS ADSORBED OXYGEN IMPORTANT?

In any study involving reactions at the surface of a metal it is important to know the properties of the surface. As discussed above, metals readily oxidize. Therefore, it is important to know whether the metal surface is actually an oxide surface or a bare metal surface. If the surface is an oxide, the properties of the oxide will govern reactions at the surface.

As one example, we may consider the oxygen reduction reaction at platinum. This reaction may either occur on a bare platinum surface or on a platinum oxide surface. The mechanism of oxygen reduction is different on the two surfaces as the Tafel slope is RT/F at bare platinum and $2RT/F$ at platinum oxide. Due to the slow kinetics of reduction of platinum oxide, it is possible to determine the kinetics of oxygen reduction on the oxide surface in the region in which the oxide is thermodynamically unstable [15]. Hence, one can get very different results in the same potential region, depending on whether the surface is platinum or platinum oxide.

Oxygen-containing species are also important as intermediates in reactions. As one example, the oxidation of ethylene at platinum may be considered. The rate-determining step has been shown to be [16]

$$H_2O + Pt \rightarrow PtOH + H^+ + e^-$$

The free energy of formation of the PtOH species is of major importance in the reaction. With a different metal, the free energy of formation of the equivalent species would be different, so that the rate of the step would be different.

Also, in the oxygen evolution dissolution reaction, oxygen-containing species must be intermediates. The free energy of bonding of these species to the substrate are a major factor in determining what path the reaction will follow.

The stability of metals is often determined by the properties of the metal oxides. Very active metals, such as aluminium and magnesium, which have adherent nonconducting oxides, behave quite inertly. The metals with conducting oxides, such as iron and nickel, become passive and behave inertly under many conditions.

The natures of the oxide and chemisorbed oxygen species thus play a very important role in determing the usefulness of metals as electrodes.

REFERENCES

1. Quinn and Roberts, *Trans. Faraday Soc.* **60**, 899 (1964).
2. Tucker, *J. Appl. Phys.* **35**, 1897 (1964).
3. Tucker, paper presented at the 150th American Chemical Society Meeting, Atlantic City, New Jersey, September 14, 1965.
4. Farnsworth, paper presented at the 150th American Chemical Society Meeting, Atlantic City, New Jersey, September 14, 1965.
5. Hayward and Trapnell, *Chemisorption*, 2nd ed., Butterworth, Washington, D. C., p. 176 (1964).
6. Young, *Anodic Oxide Films*, Academic Press, New York, p. 13 (1961).
7. Cabrera and Mott, *Rep. Progr. Phys.* **12**, 163 (1949).
8. Uhlig, *Acta Met.* **4**, 541 (1956).
9. Torrey and Whitmer, *Crystal Rectifiers*, McGraw-Hill, New York, pp. 23, 80–84 (1948).
10. Lanyon and Trapnell, *Proc. Roy. Soc.* **A227**, 387 (1955).
11. Feldberg, Enke, and Bricker, *J. Electrochem. Soc.* **110**, 826 (1963).
12. Sato and Cohen, *J. Electrochem. Soc.* **111**, 512 (1964).
13. Will and Knorr, *Z. Electrochem.* **64**, 270 (1960).
14. Andruschenko and Shishakov, *Z. Fiz. Khim.* **33**, 554 (1959).
15. Damjanovic and Bockris, *Electrochim. Acta* **11**, 376 (1966).
16. Wroblowa, Piersma, Bockris, *J. Electroanal. Chem.* **6**, 401 (1963).

Chapter 5

The Potential of Zero Charge

S. D. Argade and E. Gileadi

1. INTRODUCTION

1.1. What Is the Potential of Zero Charge?

When a metal is immersed in a solution of an electrolyte, a double layer is set up at the interface, such that there can be an excess charge on the metal side of the interface and an ionic atmosphere with a net excess of one kind of ions in the solution side of the double layer, to maintain electroneutrality over all of the system. The charge on the metal per unit area is given by

$$q^m = -F \sum_i \Gamma_i z_i \tag{1}$$

where the summation is limited to the solution side of the double layer; F is the Faraday; z_i is the charge of the ionic species i, including the sign; and Γ_i is the surface excess of that component in the interphase, in moles per unit area of the interface. The charge on the solution side of the double layer q^s is equal and opposite in sign to the charge on the metal ($q^m = -q^s$).

Consider an ideally polarized electrode, i.e., an interface across which charge transfer cannot occur. The Gibbs adsorption isotherm at constant temperature and pressure can be written as

$$d\sigma = -q^m dE + \sum_i \Gamma_i d\mu_i \tag{2}$$

and hence the excess charge density on the metal is

$$\left(\frac{\partial \sigma}{\partial E} \right)_{T,P,\mu_i} = -q^m \tag{3}$$

where μ_i is the chemical potential of the species i; E is the electrode potential; and σ is the interfacial tension. In principle, one can determine the interfacial tension as a function of the electrode potential with respect to some suitable reference electrode. The potential at which the derivative $(\partial\sigma/\partial E)_{T,P,\mu_i}$ is zero corresponds to the situation where the metallic side of the double layer has no excess charge. This potential is defined as the potential of zero charge. It will be shown below that the potential of zero charge is not a unique property of the metal, but depends on the detailed composition of the whole system. Antropov has suggested [1] that a distinction should be made between the null point of metals and the potential of zero charge. The former is meant to be a property of the metal, while the latter would only be defined for a system as a whole. The null point of a metal would be defined as its potential of zero charge in a dilute aqueous solution, in the absence of specific adsorption. While the concept of the null point of metals is useful in bringing to light the dependence of the potential of zero charge on all the components of an electrochemical system, it cannot be considered a unique property of the metal. First, specific adsorption of the solvent always occurs, and the null point hence may depend on the solvent. Second, the absorption of gases in the metal may effect its null potential. This has recently been verified [2] in the case of absorption of hydrogen in Pt and may be expected to occur with other metals and other gases (e.g., hydrogen in Fe, Al, Ti, Ni; oxygen in Ag; nitrogen in Fe; etc.).

1.2. The Rational Scale of Potential

The potential of zero charge offers itself as a natural reference point. This point of view was originally put forward by Frumkin [3] and applied by Grahame [4] to devise a rational scale of potential and was stressed further by Antropov [5]. The rational scale of potentials is referred to the potential of zero charge as zero. Thus, measured on any other scale, it will be given by

$$\bar{E} = E - E_{q=0} \tag{4}$$

where $\bar{E}$ is the potential on the rational scale; E is the measured potential; and $E_{q=0}$ is the potential of zero charge on the same scale. The potential expressed on the rational scale gives an approximate idea regarding the charge on the electrode. This is most useful for solid electrodes, where the dependence of charge on potential cannot be determined experimentally. The rational scale is independent of the reference electrode used and depends on the fundamental electro-chemical properties of the system. It is necessary to have at least an approximate idea regarding the charge on the electrode in order to apprehend the processes taking place at the electrode–solution interface, since the charge on the metal determines to a large extent the distri-bution of ions and dipoles in the solution side of the double layer.

It has been pointed out by Parsons [6,7] that the free charge q^m on the metal rather than the measured potential should be used as an independent variable in electrode processes and in particular in adsorption studies. This was further demonstrated by Bockris, Devanathan, and Muller [8] in the case of adsorption of various aliphatic compounds on Hg. These authors found that the coverage vs. charge plots had a simple and symmetrical shape. The maximum coverage occurred between -2 and -3 μCb cm^{-2} for various com-pounds, independent of the concentration in the solution. The coverage vs. potential plots on the other hand were much more complex and unsymmetrical. The potential of maximum adsorption was found to differ somewhat for different compounds and in some cases depended also on the bulk concentration. Figure 1 shows plots of θ vs. q^m and θ vs. V for the case of electrosorption of butanol on mercury.

At slightly negative charges or at slightly negative rational potential, the water molecules at the electrode–solution interface are

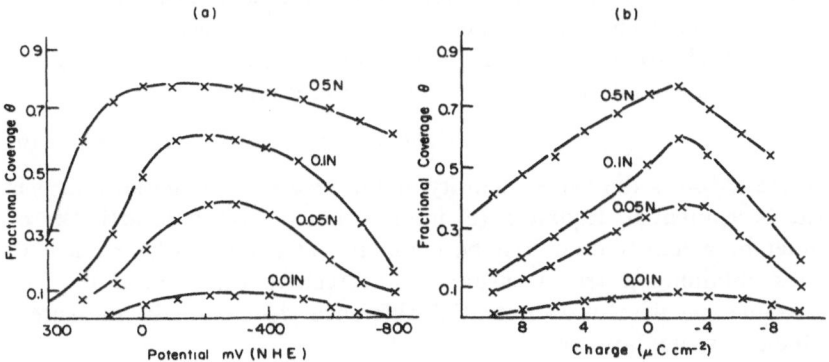

Fig. 1. Adsorption curves for butanol on Hg: (a) Coverage θ plotted against potential on the normal hydrogen scale; (b) coverage plotted against charge density on the metal. The concentration of n-butanol is as shown on the curves (from: Bockris, Devanathan, and Müller [8]).

oriented in equal numbers in an "up" and "down" position (i.e., with the two hydrogen atoms or the oxygen atom facing the metal surface, respectively). At this point the free energy of adsorption of water is a minimum and on the basis of a "competition with water" model [8-11] this corresponds to maximum adsorption of the organic species.

For charged particles, roughly the opposite sign of charge on the electrode is favorable for adsorption. Cations that are not specifically adsorbed are adsorbed when the rational potential is negative. For anions, the picture is much more complicated as anions have a larger tendency for specific adsorption. Also, whenever the interaction of the field in the double layer with some other species, charged or uncharged, becomes larger than that of the adsorbed ion, desorption occurs.

1.3. Electrode Kinetics and the Potential of Zero Charge

1.3.1. *Primary Effects.* The effect of electrode material on the rates of various types of electrode reactions has been treated recently by Frumkin [12] and by Parsons [13].

Consider a simple one-step electrochemical reaction. The reaction rate at high overpotential may be expressed in units of current density as

$$i = Fc_s \left(\frac{kT}{h} \right) \exp - \left(\frac{\Delta G^{0\ddagger}}{RT} \right) \exp \left(- \frac{\beta \Delta \phi F}{RT} \right) \qquad (5)$$

where θ_s is the concentration of reactants at the surface (more precisely at the outer Helmholtz plane); $\Delta G^{0\ddagger}$ is the chemical part of the standard electrochemical free energy of activation; and $\Delta \phi$ is the absolute metal solution potential difference, given by

$$\Delta \phi = \Delta \phi_{\text{rev}} + \eta \qquad (6)$$

where $\Delta \phi_{\text{rev}}$ is the same quantity at the reversible potential and η is the overpotential. Equation (5) implies that the catalytic activities of different metals should best be compared at a point where $\Delta \phi = 0$, thus obtaining a true measure of the relative values of $\Delta G^{0\ddagger}$. The absolute metal solution potential difference $\Delta \phi$ cannot be measured. One can write, however,

$$\Delta \phi = \Delta \Psi + \Delta \chi \qquad (7)$$

where $\Delta \Psi$ is the difference between the outer or volta potentials due to free charges and $\Delta \chi$ is the surface potential difference. It has been proposed [3] that $\Delta\chi$ is approximately independent of the metal* and since at the potential of zero charge $\Delta\Psi = 0$, this could serve as a suitable point for the comparison of the catalytic activities of different metals.

While the above argument appears superficially plausible it neglects the fact (pointed out recently by Frumkin [12]) that at equal potential $\Delta\phi$ with respect to the solution the energy levels of electrons in different metals is different. On the other hand, if two electrodes are held at the same potential φ with respect to a constant reference electrode, the energy levels of electrons in them will be equal.

Thus, if equation (5) is used to express the reaction rate, it must be remembered that $\Delta G^{0\ddagger}$ contains in it an implicit terms involving the electronic work function Φ. Equation (5) would then be rewritten as

$$i = Fc_s \left(\frac{kT}{h}\right) \exp - \left(\frac{\Delta G^{0\ddagger}}{RT}\right) \exp \left(- \frac{\beta\Phi}{RT}\right) \exp \left(- \frac{\beta\Delta\phi F}{RT}\right) \qquad (8)$$

It is noted, however, that the differences in work function between metals equal approximately the differences in the potentials of zero charge. One may write

$$\Delta\phi + \Phi = \varphi + \text{const} \qquad (9)$$

Substituting (9) into (8), one has

$$i = KFc_s \left(\frac{kT}{h}\right) \exp \left(- \frac{\Delta G^{0\ddagger}}{RT}\right) \exp \left(- \frac{\beta\varphi F}{RT}\right) \qquad (10)$$

Thus, while equation (5) is useful in clarifying the effect of the potential across the interface on the rate of charge transfer, little can be gained by comparing reaction rates at equal rational potential or at the respective potentials of zero charge (pzc), as the differences in pzc value will be canceled by the differences in work function [12,13]. Considering the physical situation, it may be seen that the rate of

---

* The assumption of $\Delta\chi \doteq$ constant may be grossly in error. This quantity is given by $\Delta\chi = \chi_m - \chi_s$. While χ_s, the surface potential of the liquid phase, may not depend substantially on the nature of the metal; little is known about the value of χ_m, the surface potential of the metalic phase. This latter quantity depends on the degree to which the electronic cloud of the metal extends beyond the surface nuclei, and there is no reason to believe *a priori* that this would be the same for all metals.

† The electronic work function may effect the reaction rate also indirectly by affecting the heats of adsorption of reactants, intermediates, and products. This point will be discussed later.

charge transfer depends on the energy of taking an electron from the metal to the solution or vice versa. This depends on the energy level of the electrons in the metal and on the total metal solution potential difference. The latter two quantities are interdependent, and at a constant potential φ with respect to a reference electrode, their sum is a constant and independent of the metal.

1.3.2. *Secondary Effects.* The discussion in the previous section should not be interpreted to imply that the position of the point of zero charge has no effect on electrode kinetics. The effects to be expected are associated with adsorption phenomena or, more generally, with the concentrations and energy levels of reactants, products, and intermediates in electrode reactions. These effects are of great importance and may cause changes of several orders of magnitude in the reaction rates. They are termed here "secondary" only in the sense that they do not affect the energy of charge transfer directly.

Equations (5) and (10) include the concentration c_s of reactants on the surface. This has been related by Frumkin [14] to the bulk concentration c_b as

$$c_s = c_b \exp\left(-\frac{\phi_2 F}{RT}\right)$$

where ϕ_2 is the diffuse layer potential or the potential at the outer Helmholtz plane, defined as the plane of closest approach of hydrated ions. The potential ϕ_2 depends primarily on the charge on the electrode, and hence on the rational potential, and on the bulk concentration c_b. At low concentrations, the variation in ϕ_2 may be as much as 20% of the variation in $\Delta\phi$, causing a change of several orders of magnitude in c_s at constant c_b. At high solution concentration, ϕ_2 and its variation with $\Delta\phi$ are much smaller and are usually neglected. It has been shown, recently, however, by Delahay [15] that this may not be justified, particularly at large values of the rational potential.

The heat of adsorption of intermediates formed in, e.g., the hydrogen or oxygen evolution reaction depend on the work function of the metal [16,17] and hence also indirectly on the potential of zero charge. When an adsorption step is rate determining (e.g., H_3O^+ discharge on Hg), a higher energy of adsorption will give rise to higher reaction rates and vice versa.

The specific adsorption of ions depends on the charge on the metal and hence also on the rational potential. This may affect the reaction rate in two ways. Directly, if the adsorbing ion is a reactant in the reaction, and indirectly, in that the adsorption of ions may affect the accessibility of the surface to other reacting species.

The effect of the potential of zero charge on the adsorption of neutral organic molecules is discussed in the next section. Suffice it to say here that since the extent of adsorption depends on the rational potential, the rates of electrochemical reactions will be affected both if the organic molecule acts as a reactant and when it serves as an inhibitor for other reactions.

1.4. Dependence of Adsorption on the Potential of Zero Charge

The predominant importance of the potential of zero charge in determining the adsorption behavior of ions and charged molecules on electrode surfaces is generally accepted and the advantage of the use of the rational potential to describe such phenomena is recognized, even though some controversy regarding the theoretical interpretation of the potential dependence of organic adsorption still exists [18-20]. The basic variable in the description of adsorption phenomena should preferably be the charge density q^m on the metal [6,7,8,18]. However, as pointed out above, this quantity is not usually accessible in the case of solid electrode and the rational scale of potential, which serves as an *approximate* measure of q^m, can be used instead. The dependence of the electrosorption of neutral molecules on potential is roughly "bell-shaped" with maximum adsorption occurring near, but not quite at, the potential of zero charge. A solvent-centered viewpoint to account for this behavior has been developed by Bockris and co-workers in recent years [8-11,21-23]. This depends fundamentally on viewing electrosorption as a replacement reaction (i.e., competition between solvent and solute for surface sites), and the dependence of the free energy of adsorption of the dipolar solvent on the field F in the double layer, given by

$$F = \frac{4\pi q^m}{\epsilon} \tag{11}$$

At the potential of zero charge the field due to free charge on the electrode is naturally zero, and the free energy of adsorption of the dipolar solvent is a minimum (in absolute value), giving rise to maximum organic adsorption. The fact that the adsorption peak does not coincide with the potential of zero charge in most cases is due to the slightly preferred orientation of the solvent molecules on the surface at zero field strength [9-11,21-23].

2. ASPECTS OF THE POTENTIAL OF ZERO CHARGE DEPENDENT ON THE METAL

2.1. The Relation Between Potential of Zero Charge, Contact Potential Difference, and Work Function

It was pointed out by Frumkin and Gorodetskaya [24] that the difference in the potential of zero charge of two metals was approximately equal to their contact potential difference $^1\Delta^2V$ is

$$^1\Delta^2V \doteq E^1_{q=0} - E^2_{q=0} \tag{12}$$

where the superscripts refer to two different metals. The deviations from equation (12) were attributed by these authors to the differences in the orientation of water dipoles at the interfaces.

Vasenin [25] made use of Frumkin's relationship [equation (12)] to correlate the electronic work function with the potential of zero charge of metals. The contact potential difference is simply equal to to the difference in work functions,

$$^1\Delta^2V = \Phi^1 - \Phi^2 \tag{13}$$

Hence,

$$E^2_{q=0} \doteq E^1_{q=0} - (\Phi^1 - \Phi^2) \tag{14}$$

The approximate nature of equation (14) may be eliminated empirically by introducing a parameter b; thus,

$$E^2_{q=0} = E^1_{q=0} - b(\Phi_1 - \Phi_2) \tag{15}$$

Since the potential of zero charge is known accurately at least for one metal (Hg), equation (15) can again be rewritten in the form

$$E_{q=0} = a + b\Phi \tag{16}$$

where a and b are empirical constants. On the basis of the available experimental data (Fig. 2), Vasenin chose the values of $a = -4.25\,\mathrm{eV}$ and $b = 0.86$. The fact that $b < 1$ was assumed to indicate that there was a specific interaction between the water dipoles and the surface layer of the metallic phase, and the contribution of this effect to the observed potential of zero charge increased with increasing work function of the metal.

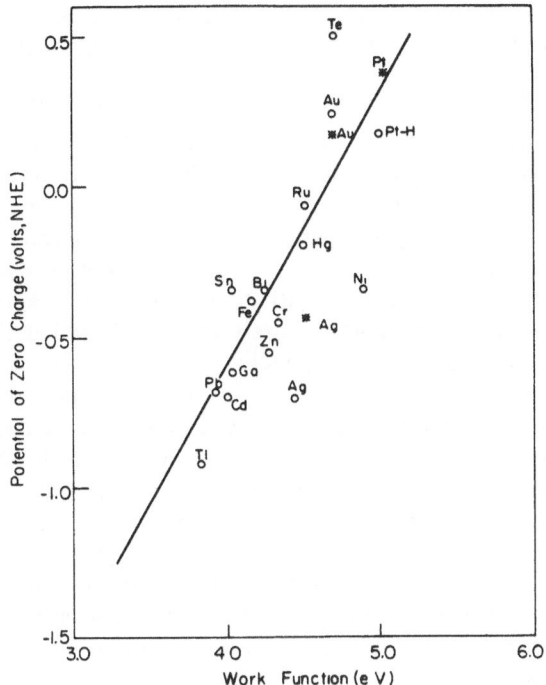

Fig. 2. Potential of zero charge plotted against the work
function; * indicates pzc values obtained in the electrochemistry laboratory, University of Pennsylvania.

Thus, an empirical relationship between $E_{q=0}$ and Φ was established.
However, the numerical values of the parameters a and b are rather
uncertain, due to the difficulty in the measurement of both experimental
quantities involved.

2.2. Physicochemical Properties of the Metal

Ukshe and Levin [26] have pointed out that the potential of zero
charge is made up of two components, one due to the bulk properties of
the metal and the other due to its surface properties. In attempting
to establish a dependence of the potential of zero charge upon the
properties of the metal, the bulk component of $E_{q=0}$ rather than its
experimental value should be considered. The difference between these
quantities may sometimes be substantial but is rarely known accurately.

A relationship between the potential of zero charge and the density and atomic weight of a metal has been suggested [26], namely,

$$(Dn/A)^{1/3} = 0.565 + 0.115 \sin 0.6(E_{q=0} + 0.28) \tag{17}$$

where D is the density; A is the atomic weight; and n is the number of free electrons in the metal per nucleus. The quantity on the left-hand side of equation (17) is the free electron density in the metal. It should decrease with increasing temperature, due to the decrease in density. Present experimental data are, however, not accurate enough to test whether the variation of $E_{q=0}$ with temperature is in agreement with this equation. Equation (17) is purely empirical, and it is claimed to be useful in the range of $E_{q=0} = 0.6$ to -1.1 V (NHE).

Frumkin has shown [28] that the metal solution potential difference at the potential of zero charge ($^{m}\Delta^{s}\phi_{q=0}$) cannot be taken equal to zero. This was confirmed by experiments [28,29] in which the potential difference between a mercury electrode at the potential of zero charge and a 1 N NaCl solution was estimated to be 0.33 V. To a first approximation, $^{m}\Delta^{s}\phi_{q=0}$ may be considered a constant [30].

The metal solution potential difference for an electrode at equilibrium with a 1 N solution of its ions is determined by the free energy changes involved in transferring the ions from the metal lattice to the solution:

$$^{m}\Delta^{s}\phi_{\text{rev}} = \frac{U_m - U_s}{ze_0} \tag{18}$$

where U_m and U_s are the free energies of the ions in the metal lattice and in solution, respectively. From a simple thermodynamic cycle one has

$$U_m = \Delta G_s + \Delta G_i - ze_0\Phi \tag{19}$$

and

$$U_s = \Delta G_h + ze_0\chi_s \tag{20}$$

where the subscripts s, i, and h refer to free energies of sublimation, ionization, and hydration, respectively; Φ is the work function; and χ_s is the potential drop across the solution–vacuum interface.

Consider now a cell made of two identical electrodes in a 1 N solution of their ions, one electrode at the reversible potential, the other at its potential of zero charge. The cell potential, which is simply the reversible potential on the rational scale $\bar{E}_{\text{rev}}$, is given by

$$\bar{E}_{\text{rev}} = E_{\text{rev}} - E_{q=0} = {}^{m}\Delta^{s}\phi_{\text{rev}} - {}^{m}\Delta^{s}\phi_{q=0} \tag{21}$$

Combining equations (18) to (21), one has

$$\bar{E}_{rev} = \frac{\Delta G_s + \Delta G_i - \Delta G_h}{ze_0} - \chi_s - \Phi - {}^m\Delta^s\phi_{q=0} \qquad (22)$$

Thus, a linear relationship between $\bar{E}_{rev}$ and the work functions Φ is derived. Equation (22) cannot be tested experimentally, because some of the quantities contained in it cannot be measured. For the purpose of numerical calculation, Ukshe and Levin [26] used the following approximate form of equation (22):

$$\bar{E}_{rev} = \frac{\Delta H_s + \Delta H_i - \Delta H_h - T\Delta S}{ze_0} - \chi_s - \Phi - {}^m\Delta^s\phi_{q=0} \qquad (23)$$

where ΔS is the entropy change involved in transferring an ion from the metal lattice to the solution and the ΔH's represent enthalpy changes for the respective processes. Table I gives the values for the quantities involved in equation (23) for a number of metals. If one designates

$$-ze_0\Phi + \Delta H_s + \Delta H_i - \Delta H_h - T\Delta S - \sum G \qquad (24)$$

then

$$\bar{E}_{rev} = \frac{1}{(ze_0)} \sum G - K \qquad (25)$$

where

$$K = \chi_s + {}^m\Delta^s\phi_{q=0} \qquad (26)$$

The nearly constant value of the parameter K shown in Table I for all metals, except Ga and Al, can serve as an indication for the validity of equation (25) and the assumptions used in deriving this equation.

Equation (23) shows that the reversible potential on the rational potential scale $\bar{E}_{rev}$ is proportional to the work function of the metal, hence the potential of zero charge is also proportional to Φ. This can be more clearly seen by considering the emf of a cell composed of an arbitrary electrode (i) at the potential of zero charge and another one (k) at its reversible potential.

$$E = E^k_{rev} - E^i_{q=0} = {}^k\Delta^s\phi_{rev} + {}^i\Delta^k\phi + {}^s\Delta^i\phi \qquad (27)$$

where ${}^i\Delta^k\phi$ represents the inner potential differences between the phases i and k. Now, taking ${}^s\Delta^i\phi_{q=0} = \text{const}$, and since

$${}^i\Delta^k\phi = \Phi_i - \Phi_K \qquad (28)$$

Table I. Physicochemical Characteristics of the Metals and the Constant in Equation (25)

Metal	z	ΔH_s, kcal/mole	ΔH_i, kcal/mole	ΔS, cal/deg	$T\Delta S$, kcal/mole	ΔH_λ, kcal/mole	Φ, eV	$ze_0\Phi$, kcal/mole	$(1/ze_0)\Sigma G$, eV	E_{rev}, V	$E_{q=0}$, V	$\bar{E}_{rev}$, V	K, V
Ag	1	69.1	174.0	7.46	2.2	116.5	4.73	109.0	0.67	+0.80	−0.05	0.85	−0.18
Cu	2	81.5	643.2	−31.6	−9.4	506.5	4.80	220.8	0.15	+0.34	−0.04	0.38	−0.23
Zn	2	31.2	628.2	−35.4	−10.5	492.5	4.25	196.0	−0.76	−0.76	−0.63	−0.13	−0.27
Cd	2	26.8	594.7	−26.9	−8.0	436.0	4.01	185.0	0.19	−0.40	−0.90	0.50	−0.31
Hg	2	14.5	670.3	−23.9	−8.1	442.1	4.53	209.0	0.88	+0.86	−0.21	1.07	−0.19
Pb	2	46.5	515.5	−10.4	−3.1	359.1	4.15	191.4	0.32	−0.13	−0.69	0.56	−0.24
Sn	2	78.0	503.6	−18.2	−5.4	384.0	4.38	202.0	0.02	−0.14	−0.35	0.21	−0.19
Al	3	67.5	1223.6	−81.7	−24.3	1102.8	4.25	294.0	−1.18	−1.67	−0.50	−1.17	0.01
Ga	3	52.0	1317.0	−93.2	−27.8	1098.0	4.20	290.5	0.11	−0.52	−0.60	0.08	0.03
Tl	3	42.8	1293.2	−57.4	−17.1	1003.0	3.68	254.0	1.39	+0.72	−0.80	1.52	−0.13
Bi	3	47.8	1126.0	−35.0	−10.1	850.0	4.40	304.5	0.42	+0.23	−0.40	0.63	−0.21

one has

$$E_{q=0}^i = (E_{rev}^k - {}^k\Delta^s\phi_{rev} - \Phi_k + \text{const}) + \Phi_i \qquad (29)$$

Thus, the linearity between the potential of zero charge and the electronic work function of a metal depends on the assumption that ${}^i\Delta^s\phi_{q=0} = \text{const}$, and some deviations from this linearity may be expected to occur.

The expression in parentheses in equation (29) can be calculated for any of the metals represented in Table I and is found to be reasonably constant. Taking the average for the metals listed in Table I (excluding Ga and Tl, which fall far from the average), one has

$$E_{q=0} = \Phi - 4.78 \qquad (30)$$

It may be recalled that Vasenin [25] gave the empirical relationship

$$E_{q=0} = 0.86\Phi - 4.25 \qquad (31)$$

and explained the departure of the slope of $E_{q=0}$ vs. Φ from unity as being due to specific interactions between solvent dipoles and the surface.

It may be concluded that, to a good approximation, the potential of zero charge is linearly dependent on the work function of the metal. Equation (30) or (31) is of limited practical use because of the experimental uncertainty in the values of work function for metals. In addition, these equations apply mainly to nontransition metal elements (cf. Table I). Deviations from equation (30) may be expected because of the differences of surface potentials for different metals and the dependence of the surface potential of the solution phase (in contact with the metal) on the metal. At present, little is known about the magnitude of these quantities.

3. ASPECTS OF THE POTENTIAL OF ZERO CHARGE RELATED TO THE SYSTEM

In the previous section the relationhsip between the physical properties of the metal and the potential of zero charge were considered. Here the effects of the components on the solution side of the interface on the potential of zero charge will be discussed. The most important factors in this respect are specific adsorption of ions, organic adsorption, and the adsorption of atomic hydrogen or oxygen.

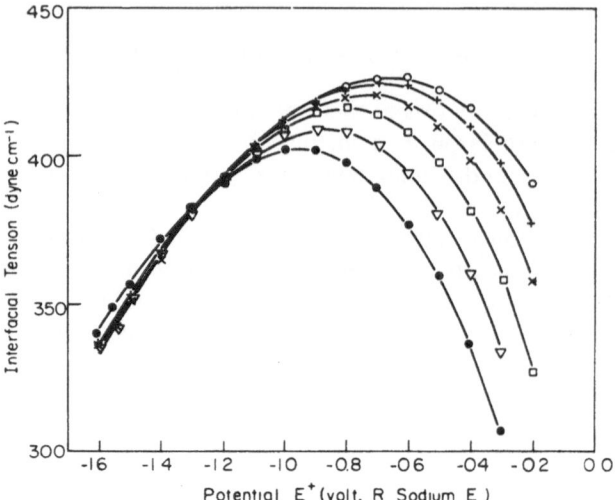

Fig. 3a. Effect of specific adsorption of anion (CNS⁻) on the electrocapillary curves. Data for this curve taken from: Z. Kovac, Ph.D. Thesis, University of Pennsylvania, Philadelphia (1965).

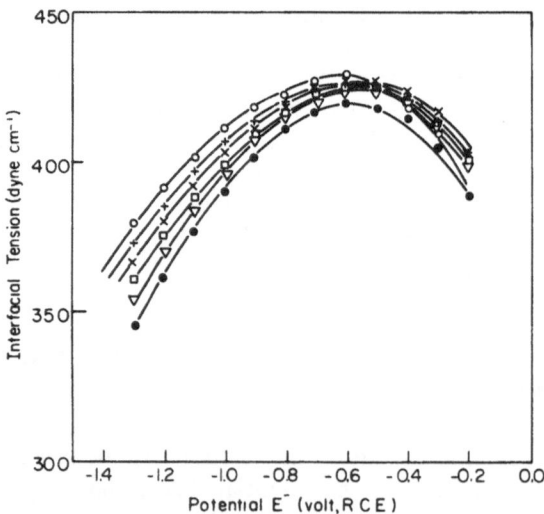

Fig. 3b. Effect of Cs⁺ cation specific adsorption on the electrocapillary curves. Data for this curve taken from Z. Kovac, Ph.D. Thesis, University of Pennsylvania, Philadelphia (1965).

3.1. Specific Adsorption of Ions

Specific adsorption of ions, particularly anions, has been studied in great detail on mercury. Figure 3a shows a typical plot of the lowering of the electrocapillary curve due to the anion adsorption. The maximum of the curve, which corresponds to the potential of zero charge, is seen to shift in the negative direction. The reverse situation is observed for specific adsorption of cations, which shifts the potential of zero charge in the anodic direction. This is shown in Fig. 3b.

Ionic adsorption has not been measured on solid electrodes with an accuracy comparable to that achieved for mercury. Qualitatively, the same kind of effect on the potential of zero charge is expected. It should be noted, however, that ionic adsorption on Pt differs markedly from that on mercury, as has been shown in a series of papers by Frumkin and his co-workers [31].

Table II shows a few of the observed values of the potential of zero charge on mercury in contact with various solutions.

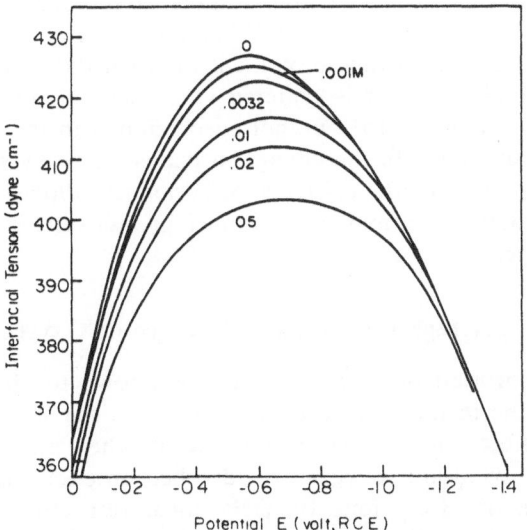

Fig. 3c. Effect of an organic compound (phenol) on the electrocapillary curves on Hg. Data from K. Müller, Ph.D. Thesis, University of Pennsylvania, Philadelphia (1965).

Table II. Variation of Potential of Zero Charge with Ionic Adsorption

Solution	$E_{q=0}$ (NHE)	Solution	$E_{q=0}$ (NHE)
1 N KOH	−0.19	1 N HCl	−0.30
1 N K$_2$CO$_3$	−0.20	1 N BaCl$_2$	−0.28
1 N K$_2$HPO$_4$	−0.21	1.0 N LaCl$_3$	−0.32
1 N Na$_2$SO$_4$	−0.20	1 N KBr	−0.37
1 N H$_2$SO$_4$	−0.23	3 N KCNS	−0.49
1 N CH$_3$COONa	−0.24	3 N KI	−0.59
1 N KNO$_3$	−0.28		

3.2. Organic Adsorption

The effect of organic adsorption of the shape of the electro-capillary curve is shown in Fig. 3c. The potential of zero charge may shift in either the cathodic or the anodic direction, depending on the compound adsorbed. The adsorbed organic molecule replaces water molecules from the interface (cf. Section 1.4 above) and thereby causes a change in the net surface dipole of the solvent and in the dielectric constant in the interface. When the molecule in the adsorbed state has no permanent dipole moment or when the functional group where the dipole moment is centered is further out toward the bulk of the solution (e.g., as for butanol adsorption with the −OH group toward the solution [18]), the change in pzc is relatively small. Strong Π-bond interaction or the existence of permanent dipoles in the inner part of the double layer may cause a much greater shift in the potential of zero charge.

3.3. Surface Coverage with Atomic Hydrogen or Oxygen

The adsorption of gases on metals is known to affect the work function of the metal, as a result of changes in the surface potential. A similar effect may be expected to occur when atomic hydrogen, OH radicals, or atomic oxygen are adsorbed on electrodes as a result of discharge of H$_3$O$^+$ ions or OH$^-$ ions, respectively. When the negative end of the dipole caused by such adsorption is outward, the work function will tend to increase and vice versa. To a first approximation, the change in work function is proportional to the partial surface coverage.

It has been shown above (cf. Section 2.2) that the potential of zero charge is approximately linearly dependent on the electronic work function. Thus, the formation of an adsorbed layer of hydrogen or an oxygen-containing species on the surface can bring about changes in the potential of zero charge of the metal.

A further complication may arise as a result of absorption of hydrogen and oxygen in the surface layers of the metal. It has recently been demonstrated experimentally [2] that the potential of zero charge of platinum can shift several hundred millivolts in the cathodic direction due to absorption of hydrogen.

3.4. pH Variation

The potential of zero charge on transition metals (e.g., Pt, Pd, Ni, Fe, Co) has been shown to vary with pH, whereas for Au, Ag, Cu, Zn, and Pb no such variation was observed [32]. Kheifets and Krasikov [32] observed that the variation of the pzc with pH occurred on metals that absorbed hydrogen, and they concluded that this was due to adsorption and/or absorption of atomic hydrogen. In a more recent study, Gileadi, Argade, and Bockris [2] have shown that the potential of zero charge for Pt occured at a potential where the coverage by adsorbed hydrogen was negligibly small. These authors also found that the dependence on pH persisted even for specimens which were thoroughly degassed and contained no absorbed hydrogen. Thus, the variation of the pzc with pH on Pt cannot be explained in terms of adsorbed or absorbed hydrogen. On the other hand, no other satisfactory explanation has so far been proposed.

4. METHODS OF DETERMINATION OF THE POTENTIAL OF ZERO CHARGE

4.1. Surface Tension Methods

4.1.1. *Electrocapillary Curves.* The electrocapillary curves are usually obtained by the capillary electrometer, which was first used by Lippmann. The interphase between a liquid metal and the solution is made in a thin capillary tube. The height of mercury required to bring the interphase to a given point in the capillary is proportional to the interfacial tension. The proportionality constant is obtained by calibration with a reference solution of known electrocapillary properties. The modern version of the electrocapillary electrometer has been described by several authors [33]. The obvious limitation of the electrocapillary measurement is that it can only be used for liquid metals, e.g., Hg, amalgams, Ga, and liquid metals in molten electrolytes [34]. Karpachev *et al.* [35] have studied Sn, Pb, Cd, Zn, Tl, Ag, Sb, Bi, Al, Ga, and Te as well as Tl, Sn, and Bi amalgams and alloys like Sn–Zn, Sn–Au, and Bi–Te in LiCi–KCl eutetic. A capillary electrometer can be substituted by a dropping mercury electrode where the drop weight method of measuring interfacial tension is used.

4.1.2. *Contact Angle Method.* Measurement of the contact angle θ between a gas bubble and a metal surface immersed in an electrolyte makes it possible to observe the interfacial tension changes. The contact angle is related to the interfacial tensions as follows:

$$\cos \theta = (\gamma_{g,m} - \gamma_{s,m})/\gamma_{g,s} \tag{32}$$

where $\gamma_{g,m}$ is the interfacial tension of gas/metal and $\gamma_{g,s}$ is that of gas/solution. Muller [36] was the first to show that contact angle changed with potential. Frumkin *et al.* [37] have determined electrocapillary curves for Pt, Ga, Zn, Ag, and Hg as well as amalgams of Tl. The values of electrocapillary maximum of mercury obtained by this method compares well with that obtained by other methods. The contact angle method is not as accurate as the electrocapillary method. Frumkin has suggested that while $\gamma_{s,m}$ varies with potential, $\gamma_{g,m}$ and $\gamma_{g,s}$ remain constant. $\gamma_{g,m}$ remains constant because of a moisture film between metal and gas. This type of behavior for ethanol–water mixtures on Hg has been confirmed by Tverdovski and Frumkin [38]. At higher concentration of ethanol, the contact angle is small and independent of potential. These authors suggested that at high ethanol concentrations the moisture film between gas and metal has bulk properties and $\gamma_{g,m}$ then varies with potential in the same way as $\gamma_{s,m}$, resulting in a constant contact angle.

This method seems to be acceptable for measuring potentials of zero charge of solid metals. Several experimental difficulties may be associated with it due to contamination of the metal surface.

4.2. Change of Surface Area

The charging current required to maintain a constant charge density on an expanding mercury drop was used by Frumkin [39] in 1923 to determine the potential of zero charge. As the area of the interface increases during the life of a drop, a current flows that is proportional to the charge density q^m on the metal and vanishes at the potential of zero charge.

This method is also useful for solid metals. While the surface area exposed to the solution changes at a fixed potential, the charging current is measured. The total charge Q on the electrode is given by

$$Q = Aq^m \tag{33}$$

where A is the area of the electrode and q^m is the charge per unit area of the electrode. Differentiating equation (33), one has

$$dQ = q^m \, dA + A \, dq^m \tag{34}$$

Hence,

$$i = \frac{dQ}{dt} = q^m \frac{dA}{dt} + AC \frac{dE}{dt} \tag{35}$$

where C is the differential capacity per unit area of the interface and E is the potential. If the area of the electrode is increased at constant potential, or the rate of change of potential with time is kept small compared to the rate of change of surface area, the second term on the right-hand side of equation (35) may be considered negligible and the charging current becomes proportional to q^m:

$$i \doteq q^m \frac{dA}{dt} \tag{36}$$

and approaches zero for any value of dA/dt as the potential approaches the potential of zero charge. There are three experimental methods available to change the surface of the metal solution interphase or create fresh surface:

1. Immersion method [40].
2. Scraping of the surface [41].
3. Crystal cleavage.

Results of the first two methods are available. Both of these measurements have been carried out in chloride solutions, where specific adsorption occurs. An inherent difficulty of the method is that if there is a current due to a Faradaic process taking place, the reversal of current one would observe may not occur at the potential of zero charge.

4.3. Capacity Measurements

This method is the most extensively used one for the measurement of the potentials of zero charge of solid metals. The capacitance of the electrode is measured by a variety of techniques as a function of potential; a-c bridge techniques or the measurement of potential differences developed across the cell by passing alternating current can be used. The former is more time-consuming but yields much more accurate data.

The potential of zero charge is discernible on the capacity-potential relation in dilute solutions of the order of $0.01 M$ or less. The capacity of an electrode can be represented as

$$1/C_{m-s} = 1/C_{m-2} + 1/C_{2-s} \tag{37}$$

C_{m-s} is the measured capacitance of metal–solution interface, which can be split up into two terms; C_{m-2}, the capacitance of the outer-Helmholtz plane, and C_{2-s}, the capacitance of the diffuse layer. Now, the diffuse layer capacitance is given by

$$C_{2-s} = |z| fA \cosh \frac{|z| f\phi_2}{2} \tag{38}$$

where

$$A = \left(\frac{RT\epsilon c_s}{2\pi}\right)^{1/2} \quad \text{and} \quad f = \frac{F}{RT} \tag{39}$$

Or for aqueous solutions with $z - z$ electrolyte at 25°C,

$$C_{2-s} = 228.5 |z| (c_s)^{1/2} \cosh(19.46 |z| \phi_2) \tag{40}$$

where C_{2-s} is in μF cm^{-2} ϕ_2 is the diffuse layer potential in volts; and c_s is the concentration in moles per liter. Near the potential of zero charge, the diffuse layer capacitance is small, and it decreases as the concentration of the electrolyte in the solution is lowered. Thus, in sufficiently dilute solution the measured capacitance is essentially equal to the diffuse layer capacitance near the potential of zero charge, provided no specific adsorption is involved. Capacitance assumes the cosh function near the potential of zero charge and one obtains a minimum that corresponds to this potential. This method has been used very successfully for the determination of the potential of zero charge. However, great difficulties may arise when there is an adsorption pseudocapacitance due to a Faradaic process taking place in the same potential range, since the adsorption pseudocapacitance is usually

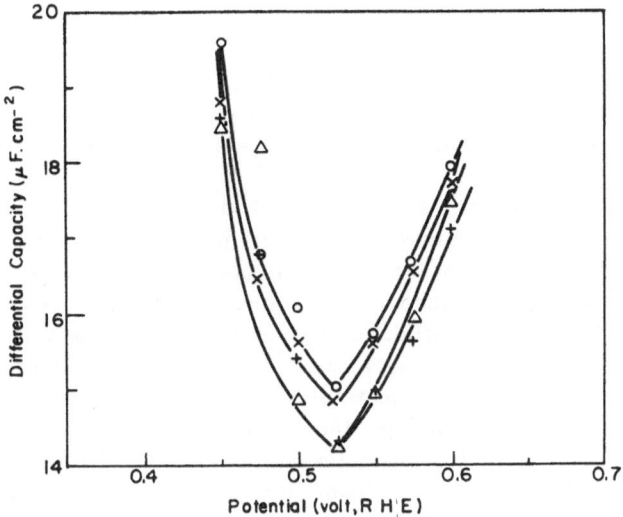

Fig. 4. Plot of double layer capacity vs. potential (RHE) for a platinum electrode in 10^{-3} N HClO$_4$ solution at various frequencies.

large compared to the ionic double layer capacitance and acts effectively in parallel with it [17,42]. Under such conditions, measurement of the capacity at a high frequency may be required to eliminate the contribution of the adsorption pseudocapacity. Even so, the existence of adsorbed species on the surface modifies the double-layer capacitance substantially and can also cause a shift in the potential of zero charge, as discussed above (cf. Section 3.3).

A typical plot of differential capacity vs. potential for a Pt electrode is shown in Fig. 4. The very small variation of capacity with frequency and the fact that the shape of the curve is independent of frequency can be taken as proof that no interference from Faradaic processes occurred under the experimental conditions employed [2].

4.4. Adsorption Methods

4.4.1. *Ionic Adsorption* [31,43]. The charge q^m on the metal electrode is given by

$$q^m = -F(z_+\Gamma_+ + z_-\Gamma_-) \qquad (41)$$

where Γ_+ and Γ_- are the surface excesses of the cations and anions, respectively, and z_+ and z_- are the corresponding charges. Balashova [43] has used a radiotracer technique to measure directly Γ_+ and Γ_- on the metal (see Frumkin et al. [31] for detailed further references). On the basis of equation (41), the charge on the electrode can then be determined as a function of potential, and the point corresponding to $q^m = 0$ is obtained.

There are numerous experimental difficulties involved in the application of this technique. The method used by Balashova et al. [43] involved measurement of the changes in concentration in solution that were very small, thus rendering the method rather insensitive. The radioactivity on the electrode after its removal from the solution was also measured, but this causes loss of potential control and the concentration of ions on the electrode may change during the transfer. In spite of these shortcomings, the approach taken by Balashova has been very fruitful in the evaluation of potentials of zero charge on solid electrodes and the understanding of double layer structure at such electrodes.

4.4.2. *Dependence of Organic Adsorption on Electrolyte Concentration.* Dahms and Green suggested a new method [44] of determining the potential of zero charge, based on measurement of the potential dependence of adsorption of neutral organic molecules and its variation with the concentration of an inert electrolyte. Since the extent of adsorption depends on the charge on the metal, a different θ vs. V curve is obtained for different electrolyte concentration, maintaining the bulk concentration of organic adsorbate constant. The curves should cross at the point of zero charge, where the charge is the same, independent of electrolyte concentration. The validity of this method was confirmed by applying it to the results of Blomgren, Bockris, and Jesch [45] for adsorption on Hg, where the pzc was known independently from electrocapillary measurement. This is shown in Fig. 5.

Inaccuracies of this method arise due to specific adsorption of the anions of the electrolyte, the change of the pzc due to organic adsorption, and the limited accuracy of the measurement of organic adsorption with presently available techniques.

4.5. Friction Methods

4.5.1. *The Pendulum Method.* Rehbinder and Wenstrom [46] claimed the existence of a relation between the interfacial tension of

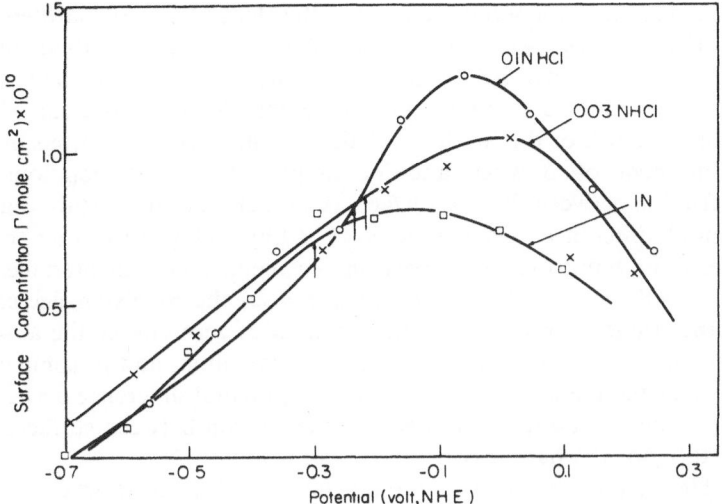

Fig. 5. Plot of surface concentration of phenol vs. potential (NHE) on mercury as a function of concentration of HCl at a constant phenol concentration of 10^{-3} N. The intersection of the two curves gives the pzc of Hg for this system in agreement with the ecm indicated by the arrows.

the metal–electrolyte system and the hardness or tensile strength of the metal surface. They measured the logarithmic decrement of the oscillations of a Herbert pendulum whose suspension is a small glass sphere about 0.5 mm diameter, resting on the metal surface covered by a drop of an electrolyte solution in which a small luggin capillary and small platinum wire (counter-electrode) were dipping. Logarithmic decrement of the oscillations was plotted vs. potential. These curves have a close similarity to the electrocapillary curves on Hg. Frumkin [47] has pointed out that if the tension on the metal is high in order to cause deformation of the surface and the fulcrum of the pendulum is made of ground glass, the decrement of the oscillations observed is due to the hardness of the metal surface. At the potential of zero charge the hardness is a maximum. If the tension at the fulcrum is reduced, however, and the ground glass ball is replaced by a smooth one, then deformation of the metal is not caused and dispersion or deformation of the surface does not determine the decrement of the oscillations, which is determined instead by the external friction as shown first by Bowden and Young [48]. Rehbinder and Wenstrom did not provide any theoretical basis for the dependence of hardness on potential or charge. Bockris and Parry-Jones [49] used a method similar to that of Rehbinder and Wenstrom. However, they have shown that friction probably determines the decrement of the oscillations of the pendulum and not the hardness. This view has been supported by the measurement of Bowden and Young [48] and by Staicopolous [50]. The friction between the two surfaces is a function of the metal–solution potential difference. On the basis of a physical picture, due to the presence of repulsion of charges on the metal–solution interface, the friction will be low. As the charge decreases the repulsion decreases, and the friction increases. At the potential of zero charge the absence of charge causes the friction to rise to a maximum, and in some cases seizure of the metal surfaces occurs. The potental difference across the metal–solution interface thus changes the friction between surfaces and probably not the hardness.

The results obtained so far by the pendulum method cannot be completely trusted because high purity conditions could not be maintained in the small amount of electrolyte at the contact between the pendulum and the metal. In particular, oxygen was not excluded. This difficulty could be overcome, in principle, by enclosing the whole apparatus in a proper atmosphere in a glove-box type arrangement. A further experimental difficulty, which may be harder to overcome, is the loss of potential control in the critical area of the fulcrum of the pendulum due to ohmic resistance in the very thin layer of solution.

4.5.2. *The Angle of Inclination Method.* The direct measurement of friction is possible as a function of potential. Bowden and Young [51] have devised a method for the measurement of static friction as a function of potential. The method in essence is as follows. A platinum wire is stretched taut in a glass cell which can be freed of gases and filled with an electrolyte. The cell is mounted on a system so that the angle of inclination of the wire can be changed. The angle of inclination α at which the cylindrical slider just starts to slide is related to the static coefficient of friction μ by the equation

$$\mu = \tan \alpha \tag{42}$$

This method for the determination of the potential of zero charge seems very promising, even though it has only been used until now by Bowden et al. [51] for the measurement of friction on Pt. It seems relatively easy to adopt for use under high purity conditions. It is also noted that the coefficient of friction is measured directly. Further, the results are not affected by slight damage to the surface of the Pt wire caused by the movement of the rider. On the other hand, no direct theoretical relationship between the potential or charge on the metal and the coefficient of friction has yet been derived.

4.6. Ultrasonic Method [52, 53]

In the acoustic field, due to the periodic distribution of charge, a periodic change in potential ($\Delta\epsilon$) is set up [52,53]. For an electrode $dq = Cd\bar{E}$, q is the charge, C is its capacity, and $\bar{E}$ is its potential in the rational scale. As the work of distribution of the charges takes place at the expense of the energy of the acoustic wave, therefore, work W is given by

$$W = \tfrac{1}{2} C_{\bar{E}}(\Delta\epsilon)^2 \tag{43}$$

The work of redistribution is proportional to the energy of acoustic waves:

$$\Delta W/\beta = \frac{I}{V} \Delta v \tag{44}$$

where V is the volume element; v and I are the velocity and intensity of the acoustic waves; and β is the coefficient of transformation of sound energy into electrical energy. Solving for $\Delta\epsilon$,

$$\Delta\epsilon = \frac{2\beta I(\Delta v)^{1/2}}{C_{\bar{E}}^2 V} \tag{45}$$

At the potential of zero charge, where capacity is a minimum and double layer is diffuse in character, one would get maximum $\Delta\epsilon$ (i.e., the ultrasonic potential). The method has been speculatively suggested by Kukoz and Kukoz [53]. No experimental work has been carried out. Another difficulty is that the objection is raised on the grounds of low Debye effect. The above authors [53] claim that the field intensities inside the hydration sheath of an ion and the double layer may differ considerably in magnitude.

4.7. Repulsion of Double Layers on Two Wires

A repulsive force between charged surfaces acts at distances of the order of the effective length of the ionic atmosphere. Thus, the liquid film will not rupture and molecular contact will not be established between two crossed metallic fibers if the external force acting on them is less than the repulsive force due to similar charges of the double layers. The energy of interaction per unit area $u(H)$ is given by

$$u(H) = N/G \tag{46}$$

N is the force of repulsion between convex surfaces at the closest approach H in the solution of the electrolyte and G is the geometric factor that depends only on the curvature and orientation of the surfaces in the region of closest approach. For cylindrical fibers of radii r_1 and r_2,

$$G = 2\pi \sqrt{r_1 r_2} \sin \theta \tag{47}$$

The apparatus used by Voropaeva, Deryagin, and Kabanov [54] is shown schematically in Fig. 6.

Each fiber could be charged to any desired potential. The fibres of platinum were 300μ in diameter. One of the fibers was attached to an elastic torsion balance of phosphor bronze with a mirror and the other to a moving part driven by a motor that could bring the

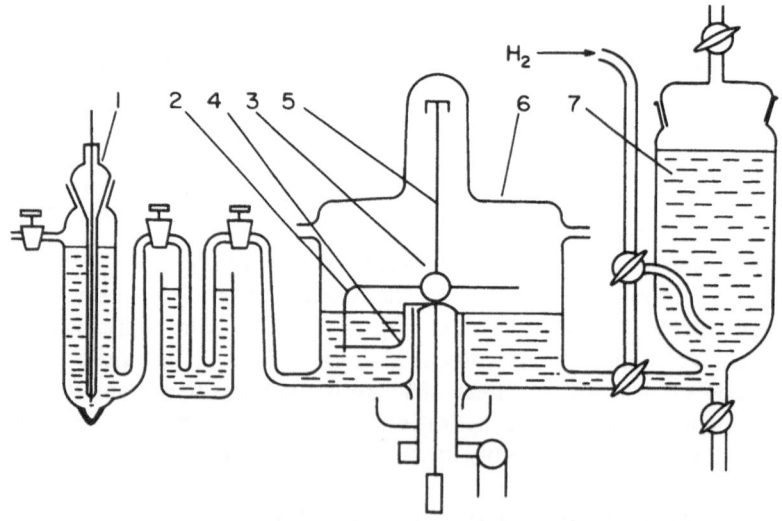

Fig. 6. Schematic diagram of the apparatus of Voropaeva, Deryagin and Kabanov [54]: (1) reference electrode; (2) Pt fiber mounted on a phosphor bronze suspension 5 with a mirror 3; (4) Pt fiber mounted on a motor drive system; (6) enclosing vessel; (7) container for the electrolyte solution.

fiber close to the other. The two fibers rotate around a common axis to avoid slippage of one past another. The whole apparatus was sealed in a vessel. The apparatus was fully automated to record in units of force, as a function of potential. When the fiber attached to the drive system was drawn close to the other fiber, a deflection was caused in the light beam shone on the mirror. An arrangement was provided to reverse the motion of the fibers at the instant of the contact between the two fibers. The fiber attached to the torsion balance keeps on moving in the same direction as the fiber approaching it until the two touch each other. The deflection caused thus gives a direct measure of the repulsive force $u(H)$. At the potential of zero charge, there being no charge on the metal, the repulsive force barrier is at its minimum.

5. CONCLUSIONS

The potential of zero charge is seen to be an important electrochemical property of the metal, but depends also markedly on the components of the solution in contact with the metal.

The rational scale of potential is defined as the difference between the measured potential and the potential of zero charge, determined with respect to the same reference electrode. The potential on the rational scale can serve as an approximate measure of the charge on the electrode. This is most useful in the case of solid electrodes where the dependence of electrode charge q^m on potential cannot be measured experimentally.

The position of the potential of zero charge (or the potential on the rational scale) affects in a primary manner the structure of the ionic double layer and the extent of adsorption of various species on the surface. Through this, the kinetics of electrode reactions are also affected, although not in a direct manner, since the differences of the potentials of zero charge between metals are canceled by the corresponding difference in electronic work function.

Many methods have been suggested in the past for the measurement of the pzc on solid electrodes. Most of these methods are difficult experimentally and yield results of low accuracy. In some cases, no solid theoretical interpretation of the phenomena involved have as yet been presented. Also, the results reported by different authors using the same or different methods are in many cases widely discrepant.

Much further experimental work is required for the determination of the potential of zero charge on solids, and techniques that allow the evaluation of the charge vs. potential relationships would be very useful in enhancing present-day understanding of the double layer at the interface between solid electrodes and solutions.

REFERENCES

1. Antropov, *The Reduced or φ Scale of Potentials and Its Application to Studies of Electrode Kinetics*, LDNTP, Leningrad, 1965.
2. Gileadi, Argade, and Bockris, *J. Phys. Chem.* **70**, 2044 (1966).
3. Frumkin, *Izv. Akad. Nauk, Otd. Khim. Nauk* 223 (1945); Frumkin, Bagotskii, Iofa, and Kabanov, *Kinetics of Electrode Processes*, Moscow University Press, USSR (1952); Frumkin, *Vestnik Mosk. Universiteta* No. 9, 37 (1952).
4. Grahame, *Chem. Rev.* **41**, 441 (1947).
5. Antropov, *Z. Fiz. Khim.* **25**, 1494 (1951); *Kinetics of Electrode Processes and Null Points of Metals*, Council of Scientific and Industrial Research, New Delhi.
6. Parsons, *Trans. Faraday Soc.* **51**, 1518 (1955).
7. Parsons, *J. Electroanal. Chem.* **7**, 136 (1964).
8. Bockris, Devanathan, and Muller, *Proc. Roy. Soc.* **A274**, 55 (1963).
9. Bockris and Swinkels, *J. Electrochem. Soc.* **111**, 736 (1964).
10. Bockris, Green, and Swinkels, *J. Electrochem. Soc.* **111**, 743 (1964).
11. Gileadi, *J. Electroanal. Chem.* **11**, 137 (1966).
12. Frumkin, *Elektrokhim.* **1**, 394 (1965).
13. Parsons, *Surface Sci.* **2**, 418 (1964).
14. Frumkin, *Z. Phys. Chem.* **164**, 121 (1933).
15. Delahay, *Double Layer and Electrode Kinetics*, Interscience, New York (1966).
16. Boudart, *J. Am. Chem. Soc.* **72**, 3566 (1952).
17. Conway and Gileadi, *Trans. Faraday Soc.* **58**, 2493 (1962).
18. Bockris, Gileadi, and Muller, *Electrochem. Acta.*, in press.
19. Frumkin, *Dokl. Akad. Nauk. Uz. SSR* **154**, 1432 (1964); Frumkin, Balashova. and Kazarinov, *J. Electrochem. Soc.* **113**, 1011 (1966).
20. Devanathan and Tilak, *Chem. Rev.* **65**, 635 (1965).
21. Dahms and Green, *J. Electrochem. Soc.* **110**, 1075 (1963).
22. Gileadi, Rubin, and Bockris, *J. Phys. Chem.* **69**, 3335 (1965).
23. Heiland, Gileadi, and Bockris, *J. Phys. Chem.* **70**, 1207 (1966).
24. Frumkin and Gorodetskaya, *Z. Physik. Chem.* **136**, 215, 415 (1928); *Z. Fiz. Khim.* **5**, 240 (1934).
25. Vasenin, *Z. Fiz. Khim.* **27**, 878 (1953) *ibid.*, **28**, 1672 (1954).
26. Ukshe and Levin, *ibid.*, **29**, 219 (1955).
27. Frumkin, *J. Chem. Phys.* **7**, 552 (1939).
28. Klein and Lange, *Z. Electrochem.* **43**, 570 (1947).
29. Latimer, Pfizer, and Slansky, *J. Chem. Phys.* **7**, 108 (1939).
30. Ershler, *Proceedings of the Conference on Electrochemistry*, Akademiya Nauk SSSR, Moscow, p. 357 (1963).

31. Frumkin, Balashova, and Kazarinov, *J. Electrochem. Soc.* **113**, 1011 (1966).
32. Kheifets and Krasikov, *Z. Fiz. Khim.* **31**, 1992 (1957).
33. Parsons, *Modern Aspects of Electrochemistry*, Vol. I, Chap. 3, Bockris and Conway, eds., Butterworth, London (1954).
34. Luggin, *Z. Physik. Chem.* **17**, 677 (1895); Hevsey and Lorenz, *ibid.*, **74**, 443 (1910).
35. Karpachev and Stromberg, *Z. Physik. Chem.* **A176**, 182; *Z. Fiz. Khim.* **10**, 739 (1937); **18**, 47 (1944); *Acta Physicochim. URSS* **12**, 523 (1940); **16**, 331 (1942); Karpachev, Kochergin, and Jordan, *Z. Fiz. Khim.* **22**, 521 (1948); Karpachev and Rodigini, *ibid.*, **23**, 453 (1949).
36. Moller, *Z. Phys. Chem.* **65**, 226 (1908).
37. Frumkin, Gorodetzkaya, Kabanov, and Nekrasov, *Phys. Z. Soviet.* **1**, 225 (1932); Gorodetzkaya and Kabanov, *ibid.* **5**, 418 (1934); Frumkin, Actualités Sci. Industr., Paris 1937, No. 373.
38. Tverdovskii and Frumkin, *Z. Fiz. Khim.* **21**, 819 (1947).
39. Frumkin, *Z. Phys. Chem.* **103**, 55 (1923).
40. Jakuszewski and Kozlowski, *Roczniki Chem.* **36**, 1873 (1962).
41. Perkins, Livingston, Anderson, and Eyring, *J. Phys. Chem.* **69**, 3329 (1965).
42. Gileadi and Conway, *Modern Aspects of Electrochemistry*, Vol. 3, Chap. 5, Bockris and Conway, eds., Butterworth, London (1964).
43. Balashova, *Z. Physik. Chem.* **207**, 340 (1957).
44. Dahms and Green, *J. Electrochem. Soc.* **110**, 466 (1963).
45. Blomgren, Bockris, and Jesch, *J. Phys. Chem.* **65**, 2000 (1961).
46. Rehbinder and Wenstrom, *Acta Physicochim. URSS* **19**, 36 (1944); *Dokl. Akad. Nauk. Uz. SSR* **68**, 329 (1949); *Z. Fiz. Khim.* **26**, 12, 12 (1952); *Z. Fiz. Khim* **26**, 1847 (1952).
47. Frumkin, *Z. Elektrochem.* **59**, 807 (1955).
48. Bowden and Young, *Research* **3**, 235 (1950).
49. Bockris and Parry-Jones, *Nature* **171**, 930 (1953).
50. Staicopoulos, *J. Electrochem. Soc.* **108**, 900 (1961).
51. Young, Ph.D. Thesis, Cambridge University (1949); Bowden and Tabor, *The Friction and Lubrication of Solids*, International Monograph in Physics by the Clarendon Press, Oxford, p. 153 (1950).
52. Yeager and Hovorka, *J. Electrochem. Soc.* **98**, 14 (1951).
53. Kukoz and Kukoz, *Z. Fiz. Khim.* **36**, 703 (1962).
54. Voropaeva, Deryagin, and Kabanov, *Izv. Akad. Nauk SSSR, Otd. Khim. Nauk* No. 2, **257** (1963).

Chapter 6

The Role of Solvents at Electrodes

K. Müller*

1. INTRODUCTION

Few of the electrodes known and used in electrochemistry are not covered with water or some other solvent. Yet in the description of electrode phenomena, the solvent has largely been disregarded, possibly because there are no methods analogous to those available for ions [1] with which to determine the solvation of electrodes quantitatively. Exceptions are the attempts that have been made to evaluate the contribution of the χ potential to the total potential drop at the metal–solution interface, but even here there is still controversy as to magnitude and even sign (cf. Chapter 5 on potentials of zero charge).

Two aspects of solvent behavior at electrodes are discussed in this chapter: (1) the dielectric behavior of the solvent layer at the electrode and (2) the role of the solvent in the adsorption of neutral solute molecules. Both are connected with the concept of a field-dependent orientation of the solvent at the interface.

The treatment will be based on experiments with polarizable electrodes, but there is no reason to assume that the considerations become invalid in the presence of a Faradaic reaction (nonpolarizable electrodes).

2. FIELD-DEPENDENT ORIENTATION OF THE SOLVENT AT THE METAL-SOLUTION INTERFACE

In 1917, Gouy [2] realized that water at the metal–solution interface plays the role of a dielectric sheet that assumes polarity under the action of the potential, and possibly even in absence of the field,

* Present address: Research Institute for Catalysis, Hokkaido, University, Sapporo, Japan.

117

giving rise to its own contribution to the potential drop. He also spoke of an orientation of its molecules by the field.

Frumkin [3-5] and Butler [6] treated the field-dependent adsorption of organic substances at electrodes as competition between them and water; however, the orientation of water was not taken as field-dependent. This idea was reborn in the early 1950's.

With respect to double-layer studies, the following equation for the interfacial capacity was obtained by Overbeek and Mackor [7-9]:

$$\frac{1}{C} = \frac{1}{K} + \frac{\partial \psi^{2-b}}{\partial q^M} + \frac{\partial \chi(q^M, c_i)}{\partial q^M} \tag{1}$$

Here, K is the constant Stern capacity for the inner part of the double layer [10]; ψ^{2-b} is the potential drop across the diffuse layer; χ is the potential caused by the oriented water at the interface, taken as a function of the charge density on the metal q^M and of the concentration of ions in solution c_i. (See also the work of Lijklema [11].)

Bockris and Potter [12] applied, independently, the concept of a variable χ potential to electrode kinetics. They explained the pH effect in hydrogen evolution from alkaline solutions on nickel by assuming discharge from water molecules whose activity depends on the electrode field so that

$$\left(\frac{\partial \eta}{\partial pH}\right)_{i_c} = \frac{RT}{F} + \frac{\partial \psi^{2-b}}{\partial pH}\left(1 - \frac{4\pi K^d \mu}{\epsilon_2 \beta F}\right) \tag{2}$$

where η is the overpotential at a given cathodic current density i_c; μ is the dipole moment of water; K^d is the integral capacity of the diffuse layer; ϵ_2 is the dielectric constant at the inner boundary of the diffuse double layer region; and β is the transfer coefficient. Equation (2) indicated correctly the order and the direction of the pH effect. A response of the orientation of water molecules to changes in the field was also implied in Bockris, Conway, Mehl, and Young's [13-15] treatment of the electric double layer as a lossy capacitor (see below).

Watts-Tobin [16] used the idea of water turning round near the point of zero charge to explain the hump that appears on capacity curves slightly anodic to the point of zero charge when he wrote

$$C \doteq \frac{K_0}{4\pi d} + K_1 \exp[f(\mu_1)] + K_2 \operatorname{sech}^2 [f'(\mu)] \tag{3}$$

where K_0 is the constant capacity of the parallel plate condenser of thickness d; K_1 is a constant referring to the capacitance arising due to adions of moment μ_1; K_2 is a constant referring to the capacitance arising due to turning water molecules of moment μ. In $f'(\mu)$, lateral dipole–dipole interactions were neglected; in $f(\mu_1)$, interactions between the adsorbed ions were also neglected.

Bockris, Devanathan, and Müller[17] showed (1) that lateral interactions in the water layer in contact with the electrode affect the rate of turning over (i.e., the change of net orientation with charge or potential) sufficiently so that no hump arises due to this on the differential capacity curves*; (2) that lateral interactions between the adsorbed ions and between these ions and their images are very considerable already at low coverages, so that the increase of the amount of adsorbed ions with charge is not exponential but undergoes an inflection, and that this feature of the adsorption isotherm of ions produces the hump observed; this was quantitatively demonstrated by Wroblowa, Kovac, and Bockris [19]; (3) that the slow reorientation of water dipoles is responsible for the main features of the adsorption behavior of neutral organic molecules, such as butanol, in aqueous solution.

The model of reorientation of water at electrodes was further developed by Bockris, Green, and Swinkels [20,21] and recently applied to methanol as solvent by Bockris, Gileadi, and Müller [22]. The idea of a field-dependent orientation of water is now generally accepted [23-29]; note particularly the statistical approach of Macdonald and Barlow [24], who treat lateral interactions as feedback effect.

3. DIELECTRIC PROPERTIES OF THE SOLVENT AT THE ELECTRODE

3.1. The Double Layer as Parallel Plate Capacitor

It had been recognized very early that the electrode–solution interface has the properties of a capacitor [30,31], and the parallel plate condenser model was used [32] without relation to molecular properties to derive correct theoretical equations for the electrokinetic phenomena. Perrin [33] put the model on a molecular basis, but the difficulty arose

---

* Cf. also Müller [83]. Dutkiewicz and Parsons [18] contend that the hump observed in formamide solution is due to reorientation of the solvent. It is desirable to test the situation here with respect to ionic adsorption isotherms.

that impossibly small values were obtained for the thickness d of the double layer (the separation of charges) when the formula for the capacity K of a parallel plate condenser was used:

$$K = \frac{\epsilon}{4\pi d} \tag{4}$$

in connection with the value of the dielectric constant ϵ for the bulk of the solution. Stern proposed that the correct value of ϵ to be used might be that of the ions [10]. But it was realized only relatively recently that a low value of ϵ must be attributed to the solvent at the electrode, although Frumkin had pointed out long ago [34] that the double layer capacities obtained in various solvents are not proportional to the respective bulk dielectric constants. Relevant work is quoted by Bockris et al. [35] for water as solvent, for which a value of approximately six leads to a consistent explanation of many experimental facts.

The general theory of dielectric dispersion was only recently applied to the polar molecules in the double layer [13–15]. The important result is that in terms of an equivalent circuit the double layer at the ideally polarizable interface behaves as a capacitor and a resistor in parallel that are inherently frequency-dependent.

3.2. Dielectric Properties of Dipolar Substances

The theory of the frequency dispersion of the permittivity (dielectric "constant") can be applied directly to the electric double layer. The relevant relations can be seen, e.g., in Böttcher's book [36], on which the presentation below is based and to which reference should be made.

The basic molecular model for the theory is that of relaxing dipoles or, more precisely, of dipoles which follow changes in the electric field E, according to a first-order rate equation so that one has for the orientation polarization P_D at any given time

$$\frac{dP_D}{dt} = -\frac{P_D}{\tau} \tag{5}$$

when the field is suddenly taken away, or

$$\frac{dP_D}{dt} = \frac{(\epsilon_s - \epsilon_\infty)E/4\pi - P_D}{\tau} \tag{6}$$

when there is a continuously varying field E present. Here, ϵ_s is the permittivity at very low frequencies (static field) where the dipoles are at all times in their equilibrium orientation with respect to the field, and ϵ_∞ is the permittivity at very high frequencies where the dipoles remain in fixed positions, i.e., are not at all oriented by the field. The region of frequencies over which the measured permittivity drops from ϵ_s to ϵ_x is called the dispersion region. In this region, the orientation polarization drops from its full contribution to zero as the frequency is increased. At the same time, a conduction mechanism is switched on (see below). Further, t is the time and τ is a characteristic relaxation time.

P_D, E, and related quantities are vector quantities that may be represented in complex notation. Using the relation between P_D, P (the total polarization), D (the dielectric displacement), and E, one can derive that

$$\epsilon = \epsilon_\infty + \frac{\epsilon_s - \epsilon_\infty}{1 + j\omega\tau} \tag{7}$$

if

$$E = E_0 \exp(j\omega t) \tag{8}$$

Equation (7) is an expression for the fact that E and D are not in phase, owing to the finite rate of reorientation of dipoles. In other terms,

$$\epsilon = \epsilon' - j\epsilon'' \tag{9}$$

A separation of real and imaginary parts of equation (7) leads to

$$\epsilon' = \epsilon_\infty + \frac{\epsilon_s - \epsilon_\infty}{1 + \omega^2\tau^2} \tag{10}$$

and

$$\epsilon'' = (\epsilon_s - \epsilon_\infty) \frac{\omega\tau}{1 + \omega^2\tau^2} \tag{11}$$

For the case that $\epsilon' \neq \epsilon_s$, the capacitor is said to have loss.

For $\tau = 10^{-6}$ sec, ϵ' and ϵ'' are shown as functions of ω in Fig. 1. In Fig. 2, ϵ'' is plotted as a function of ϵ'. In both cases, $\epsilon_s = 6$ and $\epsilon_\infty = 5$ have been used. These values were chosen to conform with various pertinent experimental results discussed below.

Debye [37] derived, for the purpose of interpretation of results obtained on impedance bridges, that

$$\epsilon' = C/C_0 \tag{12}$$

and

$$\epsilon'' = 1/\omega R C_0 \tag{13}$$

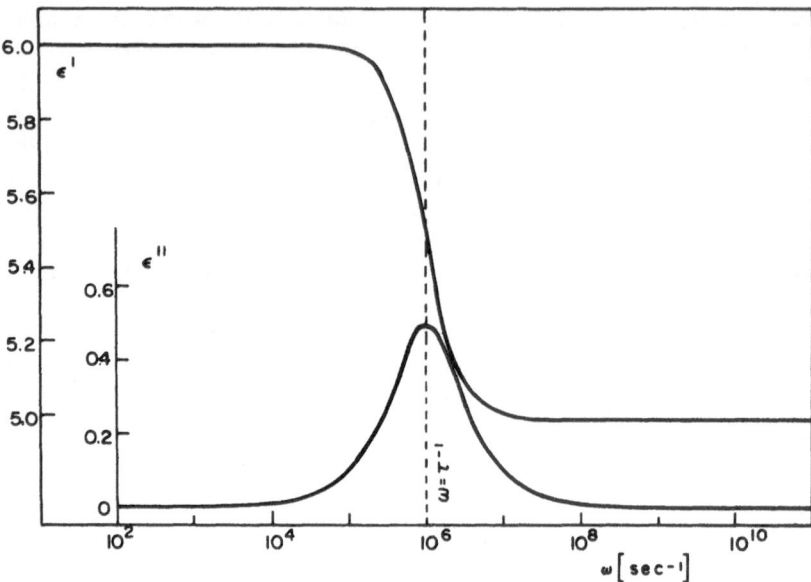

Fig. 1. Real and imaginary parts of the permittivity as a function of frequency [equations (10) and (11); $\epsilon_s = 6$, $\epsilon_\infty = 5$, $\tau = 10^{-6}$].

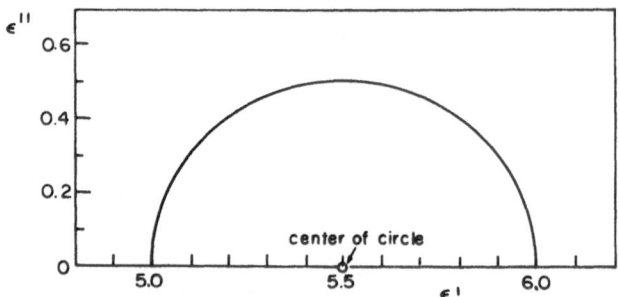

Fig. 2. Argand diagram, plotted with the values of Fig. 1.

where C_0 is the capacity of the empty condenser and C is the capacity of the filled condenser, the latter being described as having no losses but having a shunt resistor R in parallel (for the nature of the conduction current through R, see below). Introducing

$$\Delta\epsilon = \epsilon_s - \epsilon_\infty \tag{14}$$

and

$$\Delta C = C - \epsilon_x C_0 \tag{15}$$

one can obtain from equations (10) to (13) that

$$R = \frac{\tau}{\Delta \epsilon C_0}\left(1 + \frac{1}{\omega^2 \tau^2}\right) \tag{16}$$

and

$$\Delta C = \Delta \epsilon C_0 \frac{1}{1 + \omega^2 \tau^2} \tag{17}$$

For $C_0 = 6\,\mu\text{F cm}^{-2}$, values of R and ΔC were calculated and are shown in Fig. 3.

The relations developed, in particular equations (10) and (11), have permitted to represent adequately a large amount of experimental observations, even surprisingly well those on liquid and solid H_2O.

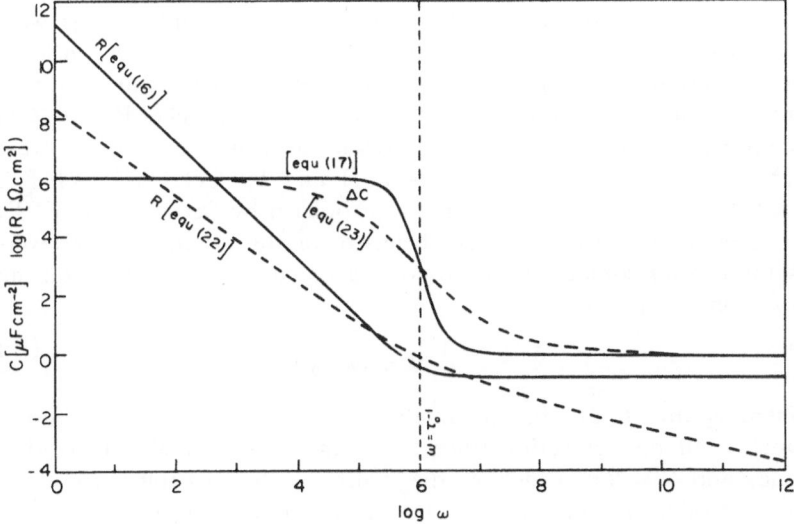

Fig. 3. Loss resistance R and capacity ΔC due to orientation polarization as functions of frequency [solid lines: equations (16) and (17), $\epsilon_s = 6$, $\epsilon_x = 5$, $\tau = 10^{-6}$, $C_0 = 6 \cdot 10^{-6}$ F cm^{-2}; no distribution of relaxation times. Broken lines: equations (22) and (23), $\epsilon_s = 6$, $\epsilon_\infty = 5$, $\tau_0 = 10^{-6}$, $C_0 = 6 \cdot 10^{-6}$ F cm^{-2}, $\beta = 0.5$; distribution of relaxation times present].

3.3. Dielectric Relaxation and Rate Theory; Distribution of Relaxation Times

As just indicated, the dielectric behavior of H_2O can be described by a single relaxation time, e.g., $\tau = 8.7 \cdot 10^{-12}$ sec at 25°C for the liquid [38] and $\tau = 5.8 \cdot 10^{-5}$ sec at -10°C for the solid [39]. By the arguments of rate theory, relaxation times can be related to the free energy of activation $\Delta G_\epsilon^{\ddagger}$ for the molecular process underlying the relaxation as [40] (cf. also Glarum [41])

$$\tau = \frac{h}{kT} \exp \frac{\Delta G_\epsilon^{\ddagger}}{RT} \tag{18}$$

Hence, a single, well-defined molecular process ought to occur under the conditions of the experiment in liquid and solid H_2O. A calculation shows that the respective values of $\Delta H_\epsilon^{\ddagger}$, the heat of activation, are nearly equal to the energies required to break one or three hydrogen bonds, respectively. $\Delta S_\epsilon^{\ddagger}$ is around 10 eu in each case. This is in agreement with other evidence on the structure of the bulk substance H_2O.

For many other pure polar substances, less simple results were obtained. These were ascribed to a distribution of relaxation times or, in the sense of equation (18), to a distribution of energies of activation around a most probable value. Equivalent (see Böttcher [36], p. 371) empirical representations for the distribution function of log τ were proposed by Fuoss and Kirkwood [42] and by Cole and Cole [43]. That of the latter authors has been introduced by Bockris et al. [13-15] to observations on the dielectric behavior of the electric double layer. Without approximations, the following equations arise. The starting equation is [43]

$$\epsilon = \epsilon_\infty + \frac{\epsilon_s - \epsilon_\infty}{1 + (j\omega\tau_0)^{1-\beta}} \tag{19}$$

where τ_0 and β are the parameters characterizing the distribution function of the relaxation times; τ_0 is the most probable relaxation time; and $0 \leqslant \beta \leqslant 1$ defines the width of the distribution. In the absence of a distribution, $\beta = 0$, and equation (7) is recovered. The wider the dispersion region, the larger the value of β.

A separation of real and imaginary parts of equation (19) leads to

$$\epsilon' = \epsilon_\infty + (\epsilon_s - \epsilon_\infty) \frac{1 + (\omega\tau_0)^{1-\beta} \sin(\beta\pi/2)}{1 + (\omega\tau_0)^{2-2\beta} + 2(\omega\tau_0)^{1-\beta} \sin(\beta\pi/2)} \tag{20}$$

and

$$\epsilon'' = (\epsilon_s - \epsilon_\infty) \frac{(\omega\tau_0)^{1-\beta} \cos(\beta\pi/2)}{1 + (\omega\tau_0)^{2-2\beta} + 2(\omega\tau_0)^{1-\beta} \sin(\beta\pi/2)} \qquad (21)$$

Instead of equations (16) and (17), one has

$$R = \frac{1}{\Delta\epsilon C_0} \left\{ \frac{1}{\cos(\beta\pi/2)} \left(\frac{\tau_0^{1-\beta}}{\omega^\beta} + \frac{\tau_0^{\beta-1}}{\omega^{2-\beta}} \right) + \frac{2}{\omega} \tan \frac{\beta\pi}{2} \right\} \qquad (22)$$

and

$$\Delta C = \Delta\epsilon C_0 \frac{1 + (\omega\tau_0)^{1-\beta} \sin(\beta\pi/2)}{1 + (\omega\tau_0)^{2-2\beta} + 2(\omega\tau_0)^{1-\beta} \sin(\beta\pi/2)} \qquad (23)$$

With the approximation $\omega \ll \tau_0^{-1}$, one obtains

$$R = \frac{\tau_0^{\beta-1}}{\Delta\epsilon C_0 \omega^{2-\beta} \cos(\beta\pi/2)} \qquad (\omega \ll \tau_0^{-1}) \qquad (24)$$

and

$$\Delta C = \Delta\epsilon C_0 [1 - (\omega\tau_0)^{1-\beta} \sin(\beta\pi/2)] \qquad (\omega \ll \tau_0^{-1}) \qquad (25)$$

Note that ΔC refers to the change in capacity between the actual frequency ω and a frequency ω_r, for which $\omega_r \gg \tau_0^{-1}$. From the experimental viewpoint, it is more convenient to refer to

$$\delta\Delta C = -\Delta\epsilon C_0(\omega\tau_0)^{1-\beta} \sin\left(\frac{\beta\pi}{2}\right) = \Delta C - \Delta\epsilon C_0 \qquad (\omega \ll \tau_0^{-1}) \qquad (26)$$

which gives the change in capacity between zero frequency (read zero logarithm of frequency) and the actual frequency.

Figure 4 shows a plot of ϵ'' against ϵ' according to equations (20) and (21), using the same parameters as before and $\beta = 0.5, \tau_0 = 10^{-6}$ sec. Figure 3 shows, in dashed lines, plots of R and ΔC according to equations (22) and (23). In the case $\beta = 0$, $\partial \ln R/\partial \ln \omega = -2$. In the presence of a distribution of relaxation times, according to equation (24),

$$-2 < \frac{\partial \ln R}{\partial \ln \omega} < -1 \qquad (0 < \beta < 1; \omega \ll \tau_0^{-1}) \qquad (27)$$

and $\partial \ln R/\partial \ln \omega$ is constant with ω under the conditions imposed on the validity of equation (27). This slope is again constant, with values between 0 and -1, under the condition $\omega \gg \tau_0^{-1}$. Figure 3 shows that the slope changes within one decade of ω on either side around $\omega = \tau_0^{-1}$.

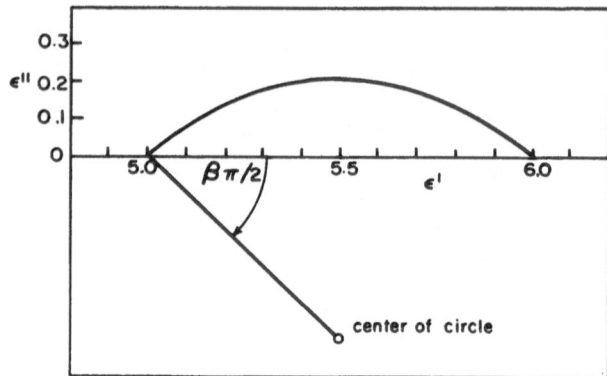

Fig. 4. Argand diagram, plotted with the parameters of Fig. 3,
for the case where a distribution of relaxation times is present
[equations (20) and (21)].

The relations presented indicate that it must be possible to detect dielectric dispersion in the electric double layer by impedance measurements at frequencies $\omega \ll \tau_0^{-1}$ if there is a distribution of relaxation times present.

3.4. Experimental Determination of the Dielectric Properties of Water in the Electric Double Layer

Impedance bridges are used widely for the measurement of the double-layer capacity. It has often been assumed that the resistive part of the impedance measured with well-polarizable electrodes is associated only with the resistance of leads and solution. This is in series with the double-layer capacitor and can be considered, within the range of frequencies of dielectric dispersion due to dipoles, to be independent of frequency so long as artifacts are excluded.

Artifactitious frequency dispersion of the double-layer impedance is difficult to exclude with solid electrodes. Measurements, therefore, were made on a dropping mercury electrode, using a thin capillary and a spherical counter electrode as introduced by Grahame [44,45]. The bridge was a Wayne-Kerr B 221 transformer ratio arm instrument; its use made it possible to exclude to a great extent artifactitious frequency variations that may arise in the various networks of capacitance and resistance standards of the bridge, in the polarizing circuit, and in the potential measuring circuit [46,47].

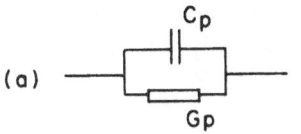

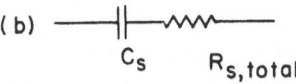

Fig. 5. Conversion of bridge readings [(a) parallel capacitance C_p and conductance G_p] into double-layer parameters [(d) parallel capacity, C_p^{dl}, and resistivity, R_p^{dl}]. (b) are the series equivalents of (a). In (c), the resistances of capillary and solution are subtracted. R_{cap} was measured directly on the bridge upon filling the cell with mercury up to the capillary, and was, typically, 35 to 40 ohms. R_{sol} was calculated (see Fig. 6) and was about 3 ohms, in the case of 6 N HCl.

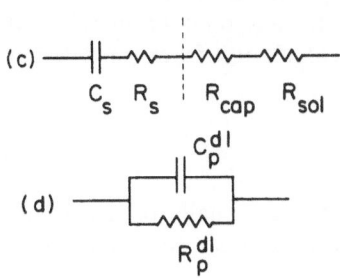

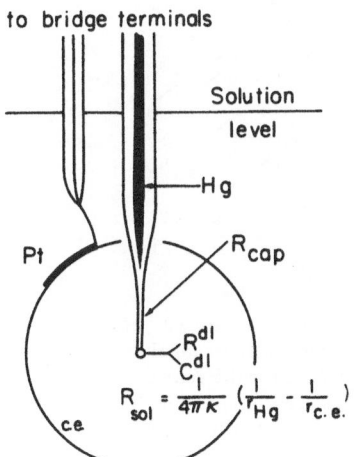

$$R_{sol} = \frac{1}{4\pi\kappa}\left(\frac{1}{r_{Hg}} - \frac{1}{r_{c.e.}}\right)$$

Fig. 6. The measuring system. The counter electrode (ce) was made of platinized platinum gauze and had, due to its large surface, a negligible impedance. In the equation, r are radii and κ is the conductivity of the solution.

The readings at bridge balance were in terms of parallel conductance G_p and capacitance C_p. Figure 5 shows how these were
converted to the desired quantities, the double-layer capacity C_p^{dl} in
μF cm^{-2}, and in parallel with it the double-layer resistivity (loss
resistance) R_p^{dl} in Ω cm^2: (a) are the measured values; (b) are their
equivalents in a series circuit; in (c), the resistance of the mercury in
the capillary (which was measured directly) and of the solution between
the drop and the counter electrode (calculated from the geometry of the
system shown in Fig. 6) are subtracted; (d) are the parallel equivalents
of what remains in (c), and these were referred to unit area. In the
calculation from (c) to (d), it was entirely sufficient to use the relations

$$R_p^{dl} \cong A/\omega^2 C_s{}^2 R_s \qquad (28)$$

and

$$C_p^{dl} \cong C_s/A \qquad (29)$$

R_s was usually of the order of several ohms at the lower frequencies.
Figure 7 shows values of R_p^{dl} obtained in the system Hg/6 N HCl$_{aq}$

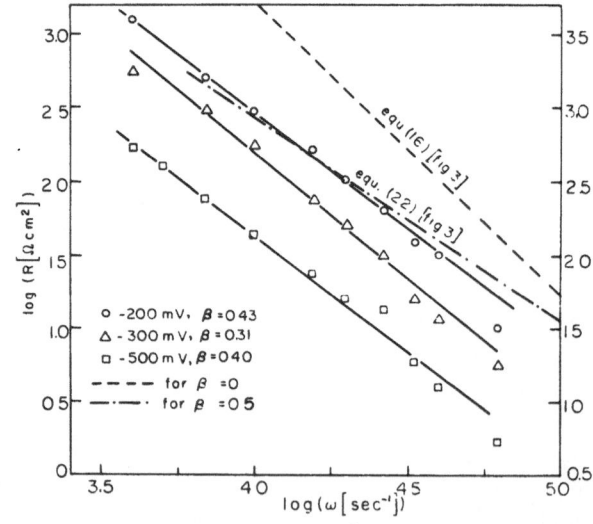

Fig. 7. Experimental values of the double layer resistivity at
the interface Hg/6 N HCl$_{aq}$, at the potentials stated (against
6 N HCl-calomel electrode). Left-hand scale for the values at
−200 mV and for the theoretical values according to Fig. 3;
right-hand scale for the values at −300 mV; left-hand scale
+1 for the values at −500 mV.

at the potentials stated, and on the same scale parts of the theoretical lines of Fig. 3. Similar results were obtained with solutions 3 N HCl$_{aq}$, 1 N HCl$_{aq}$, 0.1 N HCl$_{aq}$ and 0.1 N KCl$_{aq}$, but these were somewhat less accurate [35].

The slopes of the plots of R_p^{dl} against ω (double logarithmic) were always in accord with relation (27); most values of $\beta = 2 - \partial \ln R_p^{dl} / \partial \ln \omega$ were near 0.5. Zhilenkov [48] obtained $\beta = 0.61$ by direct dielectric studies for water adsorbed on silica gel. The observations of Epelboin and Viet [49] are also in agreement with the present results.

3.5. Theoretical Considerations and Discussion

For a discussion of the results presented in the previous section, several points must be clarified from the theoretical side: Can the results be interpreted as artifacts? If they cannot, are they consistent with a reasonable value for the most probable relaxation time τ_0? Can the value of τ_0 and the distribution of relaxation times around this value be interpreted in accordance with present views on double-layer structure? Finally, are the experimental findings on the frequency variation of the double-layer capacitance in agreement with the other evidence?

3.5.1. *Artifacts.* It is not possible to discuss here at length how surely artifacts have been excluded; a detailed error estimate [46] shows that the value of β is affected by errors of $\pm 20\%$, hence it is certain outside the experimental error limits that the observations agree with relation (27). A point that could not be controlled directly in the experiments was the penetration of electrolyte into the capillary mouth, between the mercury and the capillary walls. However, the theory of this phenomenon was well developed recently [50,51], and if such a film is of uniform thickness, relation (27) must be replaced by

$$-1 \leqslant \partial \ln R_p^{dl} / \partial \ln \omega \leqslant -0.5 \quad \text{(any frequency)} \quad (30)$$

a relation that does not conform with the experimental results. Of course, experimental conditions can be created that lead to a verification of penetration effects [51], and the theory of double-layer relaxation was criticized on similar grounds [52]; but this work does not, qualitatively or quantitatively, constitute a disproof.

It should be quite clear from the fundamental considerations that the phenomenon cannot be absent (cf. also the review of Johnson [53]). Further experimental and theoretical efforts should lead to more definite information about the position of the dispersion region and the nature of the distribution function involved, as well as about the effect of specifically adsorbed ions on this.

3.5.2. *The Value of the Mean Relaxation Time.* By comparison between Figs. 3 and 7, it is clear that $\tau_0 \leqslant 10^{-5}$ sec since the experimental curves do not indicate the existence of a bend characteristic for $\omega = \tau_0^{-1}$. On the other hand, $\epsilon_\infty \geqslant n_\infty^2$, $\epsilon_s \simeq 6$ and, therefore, from equation (15), $\Delta C \leqslant 21.5 \mu F \, cm^{-2}$ if $C = 30.7 \mu F \, cm^{-2}$ (the experimental value for $Hg/6 \, N \, HCl_{aq}$ at $-500 \, mV$ against the $6 \, N$ HCl-calomel reference electrode; this is an average value for C_p^{dl} for the potential range covered in the experiments). Using equation (24), $R_p^{dl} = 445\Omega \, cm^2$ (at $\omega = 10^4 \, sec^{-1}$) and $\beta = 0.4$ (experimental values), one obtains $\tau_0 \geqslant 10^{-7}$ sec. Thus, $\tau_0 \simeq 10^{-6}$ sec with an uncertainty of one power of ten. This indicates that double-layer water is in a condition more similar to that of ice than that of liquid bulk water (see the values of τ referred to above)—a reasonable situation.

3.5.3. *Dipole Relaxation and Double-Layer Structure.* With the value of τ_0 from the previous section, an activation energy of 9.2 (± 1.4) kcal/mole is calculated; using $\Delta S_\epsilon^\ddagger = 10$ (± 1) eu in analogy to the nearly identical values for water and ice, one obtains $\Delta H_\epsilon^\ddagger = 12.2$ (± 1.7) kcal/mole. This is considerably more than the energy (free energy, heat, entropy) of activation for transport in bulk water, e.g., the energy for formation of a hole of molecular size [40]. Therefore, it may be proposed that the process of relaxation of water adsorbed at an electrode consists of the following steps:

(*a*) Diffusion of a hole from bulk water to a position adjacent to the adsorbed water layer.

(*b*) Jump of a water molecule from an adsorbed position (in the strong double-layer field) into the empty position of the second layer (where the field is much weaker).

(*c*) Reorientation of water molecules in the second layer.

(*d*) Jump of a water molecule from the second into the adsorbed (first) layer.

In this scheme, (*b*) is the rate-determining step, and the orientation of the water molecule involved in step (*d*) is, by definition of the relaxation process, opposite to that of step (*b*).

The following additional evidence can be quoted in favor of this mechanism. First, a calculation [35] of the activation energy of step (*b*) leads to a correct order of magnitude for τ_0; $\Delta H_\epsilon^\ddagger$ was calculated to be 12.8 kcal/mole. Second, the potential-dependent (or charge-dependent) adsorption–desorption mechanism of simple organic molecules, which can be shown to depend almost entirely on the nature of the adsorption of the solvent (see below), requires the opening of sites in the first layer. In the potential region where the coverage of the electrode with those molecules changes quickly with potential, adsorption–desorption peaks are observed on the differential capacity curves (see, e.g., Melik-Gaikazyan [54] and Frumkin and Damaskin [55]). From their frequency dependence, Lorenz [56] derived an "exchange velocity of adsorption" that corresponds to a rate constant $k = 1/\tau$ of 10^5 to 10^6/sec that should (see Section 4) involve the solvent and affect adsorption equilibrium if diffusion of the organic adsorbate is fast enough. Third, it is in accord with the nature of the second layer that step (c) should be fast as compared to (b) or to a similar step in bulk solution [57]. This layer, of intermediate dielectric constant [17,58], has to provide the transition from a crowded inner layer (adsorbed layer) to the open-bulk water structure further away from the electrode.

A distribution of relaxation times logarithmically around τ_0 can be understood in terms of a distribution of energies of activation. But no satisfactory theory of this distribution exists even for the phenomena in bulk dielectrica. Qualitatively, a distribution of energies of activation can be understood with the above model (1) due to the variations in the jump distance and thus barrier height in step (b) and (2) due to changes in local structure arising from thermal fluctuations and, probably more importantly, from the movement of ions in the diffuse layer, particularly the part closest to the electrode (at the outer Helmholtz plane).

It has been shown by Glarum [41] in a statistical treatment that macroscopic observations on dielectric properties, in particular, the distribution functions involved, may indeed be interpreted directly on a molecular scale. In conclusion, therefore, agreement has been achieved in principle between the present observations on dipole relaxation in the electric double layer and current views on its structure, and the (apparently macroscopic) approach in this section is consistent with the molecular model treatment in Section 4 below.

3.5.4. *The Value of ϵ_∞ and the Mechanism of Conduction.* The range of frequencies where $\epsilon^{dl} = \epsilon_\infty$ is at present not accessible to measurements. Relatively reliable measurements [59] that have been carried out using T-bridges [56] up to $\omega = 10^7/\mathrm{sec}$ indicate a 10% lowering of the double-layer capacity in, e.g., KCl solutions. With the present theory, this would correspond to a decrease of ϵ^{dl} from ~ 6 at low frequencies to ~ 5.4 at high frequencies, in agreement with the high-frequency value of the dielectric constant for bulk water determined as 5.2 by direct dielectric studies [38,60]. The corresponding effect of frequency on capacity for $\omega \ll \tau_0^{-1}$ is very small and is readily leveled by the increase in the measured capacity due to bridge errors at high frequencies [35]; hence, studies are necessary where specially constructed bridges are used for the various frequency ranges. So far, there is tentative agreement between theory and experimentation.

The nature of the resistor shunting the capacitor in the equivalent circuit for the electric double layer described above changes with frequency. The name Maxwell resistance has been proposed [35], and this will be discussed here for the case $\beta = 0$ of Fig. 3. At $\omega \ll \tau_0^{-1}$, the resistance is due to the conduction mechanism of the ordinary Maxwell displacement current. In this region, it must be strongly frequency-dependent because the "molecular act" is reorientation of dipoles, which causes the more conduction the more often it happens; energy is dissipated, in the simplest model [37], due to the friction of the dipolar spheres with the surroundings, and the accompanying loss conductance is a consequence of the slipping of the rotating dipoles (in analogy to the slipping of an overloaded generator). At $\omega \gg \tau_0^{-1}$, the resistance is due to the movement of the same charges, but these must, for this condition, be considered as independent, quasi-free charges that move in the external field unaware of the presence of the opposite charge at the other end of the dipole[61,62]. This is similar to metallic conduction and does not depend on frequency for the purposes of the present consideration. The case where $\beta \neq 0$ corresponds to a certain mixing of these two mechanistic aspects of the conduction.

3.6. Conclusions

Experimental observations of the Maxwell resistance in the electric double layer have confirmed that this is finite and frequency-dependent, and hence the double-layer capacity must be inherently frequency-dependent. The molecular models of the relaxation process and of the double-layer structure lead to agreement between the present results and the relations derived from general dielectric theory.

4. ADSORPTIVE PROPERTIES OF THE SOLVENT AT THE ELECTRODE

Section 3 has dealt with the way of removing an oversimplification in the capacitor model of the electric double layer, in the case where there is only one dielectric present in it. The present section deals with some aspects of removing another oversimplification, consisting in the neglect of the solvent in cases where other neutral molecules become adsorbed from solution.

4.1. Gibbs Surface Excess and Adsorption of Neutral Molecules

All thermodynamically rigid derivations of the amount of substance adsorbed at an interface rest on a relation:

$$\Gamma_{i(1)} = -\left(\frac{\partial \gamma}{\partial \mu_i}\right)_{\mu_{j \neq 1,i}, T, P(\text{and } U)} \tag{31}$$

Here, $\Gamma_{i(1)}$ is the surface excess of the solution component i relative to a certain zero level of the solvent 1; γ is the interfacial tension; and μ is the chemical potential. The derivative is evaluated at the conditions of constant chemical potentials of components other than 1 and i, of constant temperature and pressure and, in the case of electrified interfaces, the potential U with respect to a certain reference electrode. The latter must be chosen to be constant (as far as changes of μ_i are concerned) if i is a neutral substance, or to respond reversibly to the concentration of the anion (cation) if the cation (anion) of the substance i is to be determined [63–69].

The true surface excess Γ_i is related to the relative one as

$$\Gamma_i = \Gamma_{i(1)} + \frac{x_i}{x_1} \Gamma_1 \tag{32}$$

(x are mole fractions), but it cannot be determined in this way since Γ_1 is not known. It can be shown in connection with equation (32) that equation (31) is based on the definition

$$\Gamma_{1(1)} \equiv 0. \tag{33}$$

Most systems of practical interest are those where $x_i/x_1 \ll 1$ and, therefore, the adsorption of the solvent has not received due attention.

Thus, values obtained for $\Gamma_{i(1)}$ are compared with the saturation value $\Gamma_{i(1)}^s$, and in this way the fractional coverage of the electrode is evaluated:

$$\theta_i = \Gamma_{i(1)}/\Gamma_{i(1)}^s \tag{34}$$

It is also taken for granted that

$$\theta_1 = 1 - \theta_i \tag{35}$$

but a value of Γ_1 cannot be evaluated from this unless arbitrary assumptions are made not only about the adsorbed layer but also about consecutive layers toward the solution. Similar assumptions are necessary to evaluate Γ_1 from the excess volume of the electric double layer [70] although, once sufficient data are available, it should in this case be possible to split experimental values of the excess volume in a way similar to that of obtaining heats of hydration of single ions.

From the reasoning behind equations (31) and (34) or (35), an unambiguous molecular picture of the electric double layer in presence of adsorption cannot be derived. In particular, adsorption isotherms must follow from this treatment that neglect the solvent. This led in the past to a direct translation of the theory of adsorption from the gas phase to that from the solution phase. The true picture, however, should be derived from a superposition of the behaviors of solvent and solute and the inclusion of interactions between all participating molecules.

This situation becomes obvious when a study of the dependence of adsorption on the electric variable (potential U or charge density on the metal q^M) is made. Thus, Frumkin [3-5,55] derived expressions for the work of replacing in the double-layer capacitor, at a given potential, slabs of dielectric consisting of solvent by those consisting of the adsorbing solute. The treatment has recently been applied widely and successfully for the interpretation of capacity data, except for the fact that the two parameters characterizing the molecular interactions and the stoichiometry at the surface remain empirical (cf. Damaskin and Tedoradze [71]). In another school, adsorption data have been used to develop a theory based mainly on the interactions of adsorbed molecules (in particular those of the solvent) with each other and with the electric field in the electric double layer, and this will be presented and discussed in the following [17,20-22].

4.2. Model of the Adsorption Process

4.2.1. *Exchange Equilibrium.* Adsorption from solution can be treated as an exchange reaction between solvent (S) and solute (A) at the surface of the electrode:

$$A_{sol} + nS_{ads} \rightleftharpoons A_{ads} + nS_{sol} \tag{36}$$

For this, Bockris and Swinkels [20] derived the applicable form of the Langmuir isotherm as

$$\frac{\theta}{(1-\theta)^n} \frac{[\theta + n(1-\theta)]^{n-1}}{n^n} = cK_0 \tag{37}$$

Here, θ refers to the coverage of the electrode surface by A; n is determined by the stoichiometry at the surface; c is the activity of A; and

$$K_0 = \exp\left(\frac{-\Delta G_a^0}{RT}\right) \tag{38}$$

contains the energy terms pertinent to reaction (36) at the charged interface. A tabulation of the function (37) was given by Swinkels [72]. To simplify calculations, cK_0 may be replaced by $c'K_e$ where c' can be used as a parameter for the effect of change of concentration of A, and K_e contains only those parts of ΔG_a^0 that are due to lateral interactions and field interactions of the adsorbed molecules (see below).

4.2.2. *The Mechanism of Adsorption.* In Section 3.5.3, the relaxation process was described in terms of a sequence of steps. Developing this scheme, one may suggest that reaction (36) follows a similar sequence, in agreement with the finite exchange rate found by Lorenz[56] and the explanation proposed above. Thus, in the presence of A and S, one can write, for the forward reaction (36) (cf. page 130):

(a) Formation of the initial system of the reaction.
   (a_1) Diffusion of a hole from bulk solution to a position adjacent to the adsorbed layer.
   (a_2) Diffusion of a molecule of A to a position adjacent to the adsorbed layer.
(b) Jump of a solvent molecule from an adsorbed position (in the strong double-layer field) into the empty position of the second layer (where the field is much weaker).
(c) Reorganization of the second layer.
   (c_1) Reorientation of solvent molecules in the second layer.

(c_2) Approach of the A molecule to the position above the free site in the first layer.

(d) Reoccupation of the position in the first layer.

(d_1) With a reoriented solvent molecule.

(d_2) With a molecule of A.

Here, (a_1) and (a_2) occur simultaneously. Steps (a_2) or (b) will be rate-determining, according to the conditions of concentration and frequency of the signal used for the observation (perturbation). Steps (c_1) and (c_2) may occur simultaneously. Step (d_1) completes the process of relaxation; (d_2) completes the process of adsorption.

For the present Section, kinetic considerations matter only in the following respect. Kinetically, (d_2) will occur most frequently upon evacuation of the site of a solvent molecule adsorbed in an unfavorable orientation of its dipole with respect to the electric field. But simultaneously there is an exchange (d_1) occurring among the solvent molecules that subsequently must lead to an equilibration between the adsorbed solvent molecules so that S in equation (36) represents the average solvent molecule. From this viewpoint, too, it must be suggested that relaxation and adsorption are really closely related; in other words, relaxation is a special case of the adsorption equilibrium (36):

$$S\!\uparrow_{\text{ads}} + S_{\text{sol}} \rightleftharpoons S\!\downarrow_{\text{ads}} + S_{\text{sol}} \tag{39}$$

Obviously, the distinction of different orientations in the bulk of the solution (where there is no electric field) makes no sense. The significance of $S\!\uparrow$ and $S\!\downarrow$ depends on the model of the adsorbed molecules, which is treated in the next section.

4.2.3. *Model of Adsorbed Molecules.* For water molecules at the interface, three models have hitherto been proposed in connection with double-layer studies. Watts-Tobin [16] understood the relatively high value of the heat of adsorption of water on mercury to mean that three "bond-forming directions" may be toward the electrode, and one toward the solution, these directions being analogous to those effective in the hydrogen-bonded, open bulk-water structure. However, calculations similar to those of Pierotti and Halsey [73] may be carried out, and these give energy values for the dispersion interaction of a sufficient order of magnitude to eliminate the need for bond formation. Devanathan and Tilak [74] suggested a position of the dipoles parallel to the interface; this model cannot account for the dielectric dispersion discussed above. Bockris *et al.* [17,20−22] allowed the solvent molecules

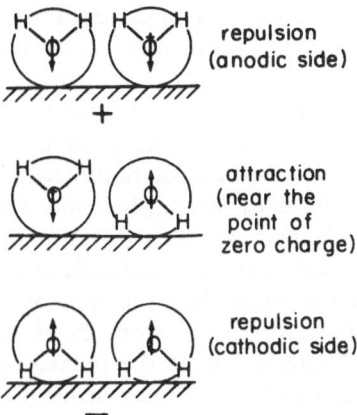

Fig. 8. Different orientations of water molecules at the metal–solution interface.

to be present, basically, in two positions,* with the dipole parallel or antiparallel to the field. Only the last model accounts for both the relaxation and the adsorption phenomena in the aqueous case and has been shown to give a satisfactory explanation of the adsorption behavior from methanolic solution.

Conforming with chemical usage (which unfortunately is the opposite of the correct physical usage), the dipoles are represented here by an arrow pointing in the direction of the negative end. Figure 8 shows preferred orientations of water for negative, negligible, and positive surface charges of the electrode. Except for the distance of the center of the dipole from the metal surface, these are symmetric, i.e., the effect of the electric field is expected to depend only on its strength, not on its direction; such a situation has often been taken [74,75] to mean that the electric part of the free energy of adsorption is a direct function of E^2 and, therefore, must be determined by the polarizability rather than the dipole moment. In fact, the average orientation of the dipoles will be a function of E so that the dependence of the electric part of the free energy of adsorption on E is more complicated than simply E^2.

A similar picture was developed for methanol as a solvent [22],

* Not identical with Watts-Tobin's, as might, to the contrary, follow from Devanathan and Tilak [74].

but in this case the dipole is not symmetric with respect to the geometry of the molecule. This leads to an inherent asymmetry of adsorption curves for neutral substances dissolved in methanol; such an asymmetry cannot arise with the polarizability of the molecule taken as the leading term in the electric part of the free energy of adsorption.

The orientation of adsorbed solute molecules is strongly affected by interactions with the bulk solvent, in a way analogous to solubility considerations, except that, due to orientation at the interface, the solubility of the various parts of a molecule will be important. Thus, butanol adsorbs on the electrode from aqueous solutions only in a position with the polar group away from the electrode [17]. In the case of n-decylamine, the interaction of the polar group is stronger with some metals than with others but always stronger than that of the polar group of butanol. This leads to a difference in orientation on various metals [20] when adsorption occurs from aqueous solution. With phenol, the π-electron interaction is sufficient to tie the molecule down to the electrode surface [76] and permit a rather strong interaction of the polar group with the electric field in the inner part of the double layer, both with water and with methanol as solvent. Finally, when butanol is adsorbed from methanol, which is a good solvent for polar as well as nonpolar substances, the position of the adsorbed solute may be flat or upright, depending on the electric field [22].

This picture of adsorbed molecules is very simple, and is certainly too simple to explain all kinetic aspects of adsorption. However, it brings out some important aspects of adsorption that cannot be derived from an analysis of adsorption isotherms. It allows, in particular, to derive the configurations of the adsorbed layer directly from molecular models (e.g., Fisher–Taylor–Hirschfelder models), and to estimate, possibly for several alternatives, the interactions between like molecules, between unlike molecules, and between molecules and the electric field. It will also be possible to include considerations about the interactions between the adsorbed molecules and the atoms of the metal surface, particularly when solid metals are involved.

4.2.4. *Molecular Interactions at the Interface.* Two types of interactions are considered: (1) between the molecules and the electric field and (2) between the molecules themselves.

The field interactions lead to terms $(\Delta\alpha/2) E^2$ and $\Delta\mu E$ where α is the polarizability of the molecules; μ is the dipole moment; E is the electric field strength; and Δ denotes differences per unit area, which may be the area occupied by one adsorbed molecule of A. In the case of polar solvents, one has the condition, up to field strengths of the order of $2 \cdot 10^7$ V/cm (corresponding to $q^M = 10 \, \mu C \, cm^{-2}$), that $(\Delta\alpha/2) \, E$ is negligible as compared with $\Delta\mu$. Further, E decreases quickly with distance from the electrode, except in very dilute solutions, so that the only field interactions arising over most of the experimental range are those of the polar groups that are in contact with the electrode. In the case of small solvent molecules, this may be the total dipole, and it may be oriented in such a way (Fig. 8) that the adsorption of S will always become stronger than that of A at increasing fields ("salting-out" action of the electric double layer).

The intermolecular forces can only be treated with some severe approximations, and it is proposed as a first solution to neglect all except those between solvent *dipoles*. Usually, the opposite is done (cf. Frumkin and Damaskin's review [55]), only interactions between adsorbed solute molecules are taken into account. However, it should be evident from the model of the electric double layer and the orientation and position of, in particular, polar parts of the adsorbed molecules that the latter approximation can only be appropriate at very high coverages. In the region of desorption, and even at coverages of A reaching, say, 2/3, the interacting groups of A will be in a part of the double layer where the dielectric constant is high, and the dipoles will have a weak electrostatic interaction. In a case such as phenol, the polar groups are in the inner layer but they are separated by a large distance when the molecules lie flat. On the other hand, solvent dipoles are always in that part of the double layer where the dielectric constant is low and, therefore, electrostatic dipole–dipole interactions are strong. Thus, the present approximation will be quite satisfactory for those cases where the molecules of A do not form condensed layers. If they do, it will be necessary to treat A–A interactions in the same way as S–S interactions.

The interactions between S and A within the adsorbed layer and also between the adsorbed molecules and molecules located toward the bulk of the solution have in the present treatment been absorbed into the constant c' (see Section 4.2.1 above); so, further, have all lateral interactions with energies decreasing more quickly with distance than third power of this, and finally all interactions between adsorbed molecules and the atoms of the electrode surface. Insofar as the dipole–dipole interactions of S were not considered as function of coverage, these approximations will not be severe.

The two main effects of the field and lateral interactions combined thus are (1) to cause a continuous change in orientation of the solvent dipoles, which is, however, much slower than could be expected by field effects alone (slow refers to the change with the electric variable) and (2) to cause a dependence of the free energy of adsorption on the electric variable both in a direct way (the term $\Delta\mu E$) and in an indirect way (change of the relative numbers of molecules of S and possibly of A adsorbed in different orientations).

4.3. The Electric Variable

When the potential is used as the (measured) quantity from which to calculate the field strength in the double layer, the following is required: The potential difference metal–solution should be known; from this, the potential drop outside the adsorbed layer (i.e., that across the diffuse layer) should be subtracted; the part due to the χ potential of polar groups outside the inner layer (which is not negligible at high coverages) must also be subtracted; and the discreteness of charge must be considered [77] since the average field at the site of an adsorbed molecule, not the average field across the double layer, should be calculated. The net quantity obtained must be divided by the relevant distance, which can easily be estimated from molecular models.

When the charge density is used to calculate the field strength in the double layer, the following is required: The potential of zero charge should be known in order to evaluate the charge density from capacity data (if electrocapillary data are used, this does not apply), and the dielectric constants in the inner part of the electric double layer should be estimated.

The difficulties connected with the former procedure are rather more fundamental than those arising in the latter case. There is, moreover, evidence that adsorption curves have a simpler shape, and depend hardly on the presence of ions in contact with the electrode (i.e., specifically adsorbed ions), when they are plotted as coverage against charge, rather than against potential [17,78]. Thus, for calculations with a molecular model of the adsorption process, the charge density must be preferred [79], and it must be required from the experimental side to determine with some accuracy the potential of zero charge in the particular system (at the given concentrations of A).

4.4. Adsorption Equations

The average energy of an adsorbed solvent molecule S can be written

$$\langle \Delta G_S \rangle = \frac{N_{S\uparrow}}{N_{ST}} \langle \Delta G_{S\uparrow} \rangle + \frac{N_{S\downarrow}}{N_{ST}} \langle \Delta G_{S\downarrow} \rangle \tag{40}$$

here,

$$\langle \Delta G_{S\uparrow} \rangle = \frac{N_{S\uparrow}}{N_{ST}} (n_c E^i_{\uparrow\uparrow})_S + \frac{N_{S\downarrow}}{N_{ST}} (n_c E^i_{\uparrow\downarrow})_S + \mu_S \cos \alpha_{S\uparrow} E + \Delta G^c_{S\uparrow} \tag{40a}$$

and a corresponding expression is written for $\langle \Delta G_{S\downarrow} \rangle$. In these equations, G denotes free energy*; $\uparrow$ and $\downarrow$ refer to orientations of the molecules; N denotes number; n_c is a two-dimensional coordination number; E^i is half the energy of interaction between the designated pair of dipoles; α is the deviation from normal to the interface of the designated dipole;

$$E = \frac{4\pi q^M}{\epsilon} \tag{41}$$

where ϵ is the dielectric constant in the inner part of the double layer, taken as six; and superscript c denotes parts of the free energy of adsorption other than those due to the dipole–field and dipole–dipole interactions.

Similarly, the average energy of an adsorbed solute molecule A can be written*

$$\langle \Delta G_A \rangle = \frac{N_{A\uparrow}}{N_{AT}} \langle \Delta G_{A\uparrow} \rangle + \frac{N_{A\downarrow}}{N_{AT}} \langle \Delta G_{A\downarrow} \rangle \tag{42}$$

here,

$$\langle \Delta G_{A\uparrow} \rangle = \mu_A \cos \alpha_{A\uparrow} E_{A\uparrow} + \Delta G^c_{A\uparrow} \tag{42a}$$

and

$$\langle \Delta G_{A\downarrow} \rangle = \mu_A \cos \alpha_{A\downarrow} E_{A\downarrow} + \Delta G^c_{A\downarrow} \tag{42b}$$

where the distinction between the respective electric field strengths is made according to the position of μ_A in the inner or outer parts of the electric double layer, and ϵ may assume different values.

---

* With the approximation that orientational contributions to the entropy, as well as those due to mixing in the adsorbed layer, sum up to a constant contained in $\Delta G_S{}^c$ and $\Delta G_A{}^c$.

In the calculations, only

$$\Delta\Delta G^c = \Delta G_\uparrow^c - \Delta G_\downarrow^c \qquad (43)$$

must be estimated; for H_2O, this has been discussed in detail [21]. Then, for given values of $n_c E^i$, α, and $\Delta\Delta G^c$, equations (40) and (42) can be solved for q^M as a function of the auxiliary variable B (i.e., B_S and B_A, respectively):

$$B = \frac{N_\uparrow - N_\downarrow}{N_T} = \frac{\exp\langle -\Delta G_\uparrow/RT\rangle - \exp\langle -\Delta G_\downarrow/RT\rangle}{\exp\langle -\Delta G_\uparrow/RT\rangle + \exp\langle -\Delta G_\downarrow/RT\rangle} \qquad (44)$$

thereupon, $\langle\Delta G_S\rangle$ and $\langle\Delta G_A\rangle$ can be calculated as functions of q^M. With

$$k_0 = \exp(\langle -\Delta G_A\rangle + n\langle\Delta G_S\rangle) \qquad (45)$$

and equation (37), the coverage can be obtained for various values of the parameter

$$\Delta G_a^c = \Delta G_A^c - n\Delta G_s^c \qquad (46)$$

which is part of the exponent in equation (45).

In this procedure, $n_c E^i$, α, and n can be estimated from molecular models of A and S, and a value for ΔG_a^c can be obtained from the value for θ_A at zero charge at the given concentration. ΔG_a^c has been related to solubility [76]. In this sense, none of the parameters is empirical, although the absolute value of $n_c E^i$ is uncertain, and it is only a rough approximation to consider it as a constant with q^M and with θ.

Calculations actually carried out using a single set of parameters [22] (cf. also Heiland et al. [80]) gave rather satisfactory agreement for adsorption on mercury when $S = H_2O$ or MeOH and $A =$ butanol or phenol. The experimental and theoretical adsorption curves for the system H_2O/phenol are shown in Fig. 9; for these calculations, it was assumed that $N_{A\uparrow} = 0$, $E^i n_c/RT = 2$, $n = 4$, $\Delta\Delta G_s^c = RT$, and $\mu_A E_A/RT = 0.42\, q^M$. The remaining larger discrepancy toward the far anodic side indicates, possibly, an increasing effect of an interaction of the π-electrons of phenol with the metal surface or an effect of strong specific adsorption on the A–S competition (reduction of N_{ST}).

4.5. Conclusions

Section 4 was concerned with an absolute calculation of the variation of adsorption from solution with the electric variable at the electrode–solution interface. This was carried out in satisfactory

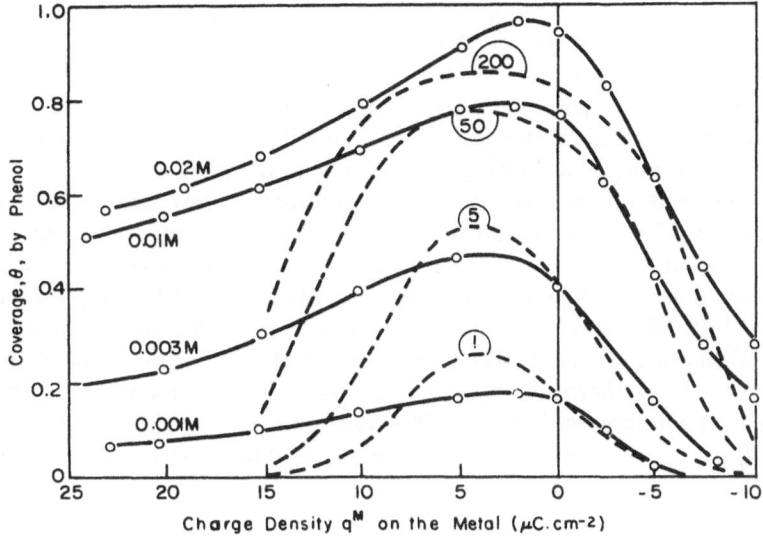

Fig. 9. Adsorption (fractional surface coverage) of phenol from aqueous solution. Solid lines: experimental values at the concentrations stated. Broken lines: values derived with equations (37), (40), (42), and (45) with the assumptions and parameters described in the text. The values given in circles are relative concentrations.

agreement with experimental results on the basis of a realistic double-layer model that can also be used to explain dielectric relaxation phenomena at the interface and is strictly based on molecular properties. Those of the solvent were found to be decisive.*

5. FINAL REMARKS

A fairly consistent picture has been reached, with some approximations, in the explanation of basic features of relaxation and adsorption at the interface. Little progress has so far been made in the quantitative determination of the amount of solvent at the interface. Some calculations are given in the work of Andersen and Bockris [81], and excess volumes at the interface were first determined by Hills and Payne [70]. This work should be developed particularly with the

---

* Cf. a similar conclusion by Andersen and Bockris [81] with respect to the adsorption of ions.

pressure variation method in connection with studies of adsorption using differential capacity data. The absolute amount of solvent in contact with the electrode might then be available from an extrapolation of a larger amount of data on interfaces of a mixed composition. Relaxation phenomena at mixed interfaces (in systems where bulk diffusion is not limiting) should add to the picture.

Little is known, so far, on the quantitative correlations between the adsorption behaviors of neutral substances and ions. Independence has been assumed as first approximation, particularly when comparing double layers at like charge densities. Conway et al. [82] have presented experimental results that indicate this approximation breaks down in solutions of high base-electrolyte concentration. Future work should center on isotherm determinations that allow evaluation of this effect separately (cf. Müller [79]) and proceed in the simultaneous analysis of the adsorption behavior of all components, charged and neutral.

NOTE ADDED IN PROOF

Recent work [84] indicates that, while the dipole–dipole interactions of the solvent are most important for the variation of adsorption with the electric variable, with which the theory presented is concerned, London-type dispersion interactions involving the adsorbed solvent molecules give rise to energy terms in the adsorption equation that are proportional to θ_A. These latter interactions of the solvent molecules may be the main contribution to the interaction parameter in Frumkin's isotherm [55] for the case of adsorption of neutral molecules.

REFERENCES

1. Conway and Bockris, *Modern Aspects of Electrochemistry*, Vol. 1, Chap. 2, Bockris, ed., Butterworth, London (1954).
2. Gouy, *Ann. Phys.* [9] **7**, 129 (1917).
3. Frumkin, *Z. Physik. Chem.* **116**, 466 (1925).
4. Frumkin, *Z. Physik* **35**, 792 (1926).
5. Frumkin and Obrucheva, *Biochem. Z.* **182**, 220 (1927).
6. Butler, *Proc. Roy. Soc. (London)* **A122**, 399 (1929).
7. Overbeek and Mackor, *Proc. 3rd Meeting Intern. Comm. Electrochem. Thermodyn. Kinet.* p. 346 (1951).
8. Mackor, *Rec. Trav. Chim.* **70**, 747 (1951).

9. Mackor, *Rec. Trav. Chim.* **70**, 763 (1951).
10. Stern, *Z. Elektrochem.* **30**, 508 (1924).
11. Lijklema, *Adsorptie van tegenionen*, Proefschrift, Utrecht (1957).
12. Bockris and Potter, *J. Chem. Phys.* **20**, 614 (1952).
13. Bockris, Mehl, Conway, and Young, *J. Chem. Phys.* **25**, 776 (1956).
14. Bockris, Mehl, and Conway, *Proceedings of the 4th Conference on Electrochemistry*, 1956, Academy of Sciences of the USSR, Moscow (1959), p. 380.
15. Bockris and Conway, *J. Chem. Phys.* **28**, 707 (1958).
16. Watts-Tobin, *Phil. Mag.* **6**, 133 (1961).
17. Bockris, Devanathan, and Müller, *Proc. Roy. Soc. (London)* **A274**, 55 (1963); *Proceedings of the 1st Australian Conference on Electrochemistry*, Pergamon Press, Oxford, p. 832 (1965).
18. Dutkiewicz and Parsons, *J. Electroanal. Chem.* **11**, 196 (1966).
19. Wroblowa, Kovac and Bockris, *Trans. Faraday Soc.* **61**, 1523 (1965).
20. Bockris and Swinkels, *J. Electrochem. Soc.* **111**, 736 (1964).
21. Bockris, Green, and Swinkels, *J. Electrochem. Soc.* **111**, 743 (1964).
22. Bockris, Gileadi, and Müller, *Electrochim. Acta*, in press.
23. Schwartz, Damaskin, and Frumkin, *Zh. Fiz. Khim.* **36**, 2419 (1962).
24. Macdonald and Barlow, *J. Chem. Phys.* **39**, 412 (1963).
25. Epelboin and Viet, *J. Chim. Phys.* **59**, 857 (1963); *Proceedings of the 1st Australian Conference on Electrochemistry*, Pergamon Press, Oxford, p. 261 (1965).
26. Johnson, *Electrochim. Acta* **9**, 653 (1964).
27. Minc and Jastrzębska, *Roczniki Chem.* **37**, 507 (1963).
28. Minc and Jastrzębska, *Electrochim. Acta* **9**, 533 (1964).
29. Frumkin, Polyanovskaya, and Grigor'ev, *Dokl. Akad. Nauk SSSR* **157**, 1455 (1964).
30. Kohlrausch, *Pogg. Ann.* **148**, 143 (1873).
31. Kohlrausch, *Pogg. Ann., Jubelband*, 290 (1874).
32. Helmholtz, *Wied. Ann.* **7**, 337 (1879).
33. Perrin, *J. Chim. Phys.* **2**, 601 (1904).
34. Frumkin, *Ergeb. Exakt. Naturw.* **7**, 235 (1928).
35. Bockris, Gileadi, and Müller, *J. Chem. Phys.* **44**, 1445 (1966).
36. Böttcher, *Theory of Electric Polarisation*, Chap. 10, Elsevier, Amsterdam (1952).
37. Debye, *Polar Molecules*, Chemical Catalog Company, New York, 1929.
38. Lane and Saxton, *Proc. Roy. Soc. (London)* **A213**, 400 (1952).
39. Auty and Cole, *J. Chem. Phys.* **20**, 1309 (1952).
40. Glasstone, Laidler, and Eyring, *Theory of Rate Processes*, Chap. 9, McGraw-Hill Book Company, New York (1941).
41. Glarum, *J. Chem. Phys.* **33**, 1371 (1960).
42. Fuoss and Kirkwood, *J. Am. Chem. Soc.* **63**, 385 (1941).
43. Cole and Cole, *J. Chem. Phys.* **9**, 341 (1941).
44. Grahame, *J. Am. Chem. Soc.* **63**, 1207 (1941).
45. Grahame, *J. Am. Chem. Soc.* **68**, 301 (1946).
46. Müller, Ph.D. Dissertation, University of Pennsylvania (1965).
47. Cahan, Staicopolus, and Müller, *Differential Impedance Measurements in Double Layer Studies*, to be published.
48. Zhilenkov, *Zh. Fiz. Khim.* **36**, 2406 (1962).

49. Epelboin and Viet, *J. Chim. Phys.* **59**, 857 (1963).
50. Grantham, Ph.D. Dissertation, Iowa State University (1962).
51. Leikis, Sevast'yanov, and Knots, *Zh. Fiz. Khim.* **38**, 1833 (1964).
52. Levie, *J. Electroanal. Chem.* **9**, 117 (1965).
53. Johnson, *Electrochim. Acta* **9**, 653 (1964).
54. Melik-Gaikazyan, *Zh. Fiz. Khim.* **26**, 560 (1952).
55. Frumkin and Damaskin, *Modern Aspects of Electrochemistry*, Vol. 3, Chap. 3, Bockris and Conway, eds., Butterworth, London and Washington, D. C. (1964).
56. Lorenz, *Z. Physik. Chem. (Frankfurt)* **26**, 424 (1960).
57. Conway, Bockris, and Linton, *J. Chem. Phys.* **24**, 834 (1956).
58. Mott, Parsons, and Watts-Tobin, *Phil. Mag.* **7**, 483 (1962).
59. Lorenz and Krüger, *Z. Physik. Chem. (Leipzig)* **221**, 231 (1962).
60. Hasted and El Sabeh, *Trans. Faraday Soc.* **49**, 1003 (1953).
61. Kauzmann, *Rev. Mod. Phys.* **14**, 12 (1942).
62. Hippel, *Dielectrics and Waves*, John Wiley & Sons, New York, p. 177 (1954).
63. Gibbs, *The Collected Works*, Longmans, Green, New York (1928); *Trans. Conn. Acad.* **3**, 343 (1878).
64. Parsons and Devanathan, *Trans. Faraday Soc.* **49**, 404, 673 (1953).
65. Guggenheim and Adam, *Proc. Roy. Soc. (London)* **A139**, 218 (1933).
66. Grahame and Whitney, *J. Am. Chem. Soc.* **64**, 1548 (1942).
67. Parsons, *Modern Aspects of Electrochemistry*, Vol. 1, Chap. 3, Bockris, ed. Butterworth, London (1954).
68. Frumkin and Iofa, *Zh. Fiz. Khim.* **13**, 931 (1939).
69. Grahame and Soderberg, *J. Chem. Phys.* **22**, 449 (1954).
70. Hills and Payne, *Trans. Faraday Soc.* **61**, 326 (1965).
71. Damaskin and Tedoradze, *Electrochim. Acta* **10**, 529 (1965).
72. Swinkels, Ph.D. Dissertation, University of Pennsylvania (1963).
73. Pierotti and Halsey, *J. Phys. Chem.* **63**, 680 (1959).
74. Devanathan and Tilak, *Chem. Rev.* **65**, 635 (1965).
75. Parsons, *J. Electroanal. Chem.* **7**, 136 (1964).
76. Blomgren, Bockris, and Jesch, *J. Phys. Chem.* **65**, 2000 (1961).
77. Devanathan, *Proc. Roy. Soc. (London)* **A264**, 133 (1961).
78. Parsons, *J. Electroanal. Chem.* **8**, 93 (1964).
79. Müller, *On the Description of Adsorption from Electrolytes*, to be published.
80. Heiland, Gileadi, and Bockris, *J. Phys. Chem.* **70**, 1207 (1966).
81. Andersen and Bockris, *Electrochim. Acta* **9**, 347 (1964).
82. Conway, Barradas, Hamilton, and Parry, *J. Electroanal. Chem.* **10**, 416 (1965).
83. Müller, *J. Res. Inst. Catalysis, Hokkaido Univ.* **14**, 224 (1967).
84. Müller, *The Frumkin Isotherm and Competitive Adsorption*, in press.

Chapter 7

Theoretical Consideration of the Double Layer

H. D. Hurwitz

1. AN ELECTROSTATIC MODEL OF THE ELECTROCHEMICAL DOUBLE LAYER

1.1. General Features of the Model

In order to represent the electrochemical double layer at a plane interface, one assumes generally that the dielectric behavior is distinctly different in two regions, as illustrated in Fig. 1.

Region I accounts for the preferential orientation of water dipoles in close contact with the metal. Such orientation arises under the coupled influence of the specific interaction with the metal surface and of

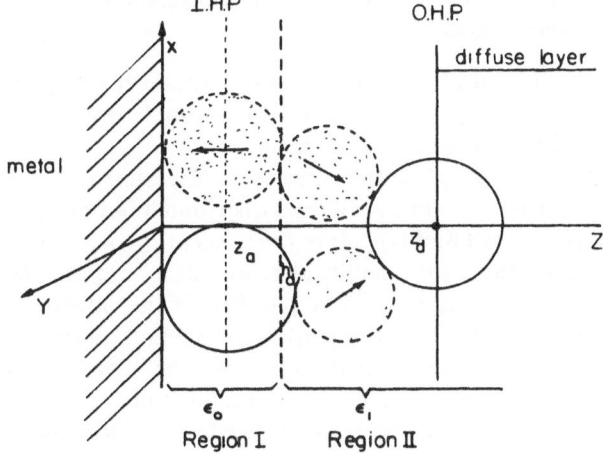

Fig. 1. The model of electrochemical double layer.

147

the effective interfacial electric field. Under certain conditions, this orientation effect may almost eliminate the orientation polarization and let the electronic and nuclei polarization of the water molecules determine the value of the local dielectric permittivity. The thickness of Region I, h_d, is of the order of the molecular dimension of one water dipole.

Region II extends from $z = h_d$. Proceeding in the direction away from the metal, the permittivity is considered to increase abruptly at $z = h_d$ to the value ϵ_1. The permittivity in this part of the double layer is considered to be smaller than the bulk permittivity due to the dielectric saturation resulting from the local electrostatic field. It will be shown later that a fall of ϵ_1 may also result whenever, in region II, there is absorption of hydrated species or of particles (like H^+ and OH^-), inducing in their vicinities a strong ordering effect on the water structure.

The outer Helmholtz plane (OHP), defined by the centers of the fully hydrated ions at their closest approach to the electrode, is located at $z = z_d$. The plane parallel to the metal surface at z_d thus constitutes the internal boundary for the diffuse part of the double layer. The centers of the specifically adsorbed ions at the metal surface are confined in a very narrow region almost coinciding with the inner Helmholtz plane (IHP) at $z = z_a$.

1.2. Distinction Between the Electrostatic Interactions in the Double Layer

It is generally admitted that the electrostatic force acting on an ion α adsorbed at r_1 in the diffuse part of the double layer ($z > z_d$) may be decomposed into several distinct terms.

1. The first of these terms consists of the sum of the coulomb potentials at r_1 arising from the charges distributed in both parts of the double layer (Regions I and II) before the adsorption of α at position r_1. Under these conditions the charges at z_a and at $z \geqslant z_d$ are assumed to be randomly spread out on planes parallel to the interface, so that the electrostatic potential at r_1, $\bar{\psi}(r_1)$ is of average nature. It should be mentioned that the contribution of the total amount of adsorbed charges, except α, results simultaneously from a direct coulombic effect and from the coulombic effect of their image charges in the electrode and in any other dielectric discontinuity of the system (like the plane $z = h_d$). The average electrostatic potential is evaluated with reference to the bulk potential, considered as zero.

2. The introduction of α at r_1 modifies the charge distribution in the surrounding. There is also a shifting of the local uniform charge density out of the volume now occupied by α. An ionic atmosphere and cavity field result, both acting on the ion α and pertaining to its activity in the double layer.

3. The ion α interacts with its own image in the metal surface. As in most electrode systems, [1] this contribution is generally included into the specific (or standard) part of the electrochemical potential.

Whenever the particle α is specifically adsorbed at z_a on the IHP, the three preceding electrostatic effects remain, yet with a comparatively larger contribution of bonds with image charges. Furthermore, a fourth interaction of coulombic nature must be introduced.

4. This accounts for the presence of an adsorbed solvent dipole layer between $z = 0$ and $z = h_d$. In the same way as for 1, one may first evaluate the electrostatic potential $\overline{\chi}(r_1)$ at r_1 (on z_a) corresponding to an ideal dipole layer of charge components uniformly distributed on two parallel plates. The value of the potential $\overline{\chi}(r)$ so obtained refers to a zero value on the bulk side of the plane h_d. The introduction of ions on the IHP affects the local distribution of dipoles, and hence modifies their electrostatic interaction with a given charge α. The latter effect has to be included in the activity of the adsorbed species.

It is our aim to limit ourself in this section to the evaluation of the average electrostatic potential $\overline{\psi}(r_1)$ and to consider the other interactions in the next sections.

1.3. A Closed Solution for the Coulomb Potential of an Isolated Charge

Taking into account the symmetry of our double-layer model (Fig. 1) with respect to an axis z normal to the electrode, it may be inferred that the electrostatic variables will be most simply expressed in a cylindrical coordinate system. The axis z is oriented toward the bulk of the solution and its origin will be taken at the metal surface. The distance $s_{12}^2 \equiv (x_2 - x_1)^2 + (y_2 - y_1)^2$ is to be defined in a plane parallel to the electrode.

As an application of the solution of Laplace's equation for cylindrical boundaries, an expression for the reciprocal of the distance r_{12} between two points 1 and 2 is readily obtained [2]:

$$\frac{1}{r_{12}} = \frac{1}{(z_{12}^2 + s_{12}^2)^{1/2}} = \int_0^\infty \exp(-kz_{12}) \, J_0(ks_{12}) \, dk \qquad (1)$$

where $J_0(ks)$ is the Bessel function of zero order. From equation (1), the potential at 2 due to a point charge i located at $1'$, $\psi_i(r_{12})$ is easily computed:

$$\Psi_i(r_{12}) = \frac{e_i}{\epsilon} \int_0^\infty \frac{\exp(-kz_{12})}{k} J_0(ks_{12})kdk \tag{2}$$

This is a Hänkel transform integral [3] and $e_i \exp(-kz)/\epsilon k$ is the Hänkel transform of order zero of the potential $\psi_i(r_{12})$.

At the planes $z = 0$, $z = h_d$, facing one of the regions defined in Fig. 1, a given fraction of the charge e_i will be imaged. Since (2) is a solution of Poisson's equation involving an isolated charge e_i, it is evident that in the presence of sheets of different permittivity near the metal surface the solution of the electrostatic problem is obtained by summing a given number of terms similar to (2), each of them representing the contribution attributable to the effect of one of the permittivity discontinuities at $z = 0$ or $z = h_d$. Furthermore, since the Poisson equation involves z and s only, we shall still have a solution if any function of k is inserted under the integral of (2). Therefore, by defining $\psi_{iI}(r_{12})$ and $\psi_{iII}(r_{12})$ as the potential at 2 due to a point charge e_i at 1, if respectively 2 is located in Regions I and II, and using the superscripts A and D if respectively the location 1 of e_i is situated in I or II, we may write, under the most general form, that

$$\Psi_{iI}^A(r_{12}) = \frac{e_i}{\epsilon_0} \int_0^\infty \left\{ \frac{\exp(-k|z_1 - z_2|)}{k} + \frac{A_0^A(k)}{k} \exp(-kz_2) \right.$$

$$\left. + \frac{B_{h_d}^A(k)}{k} \exp(kz_2) \right\} J_0(ks_{12})kdk \tag{3a}$$

$$\Psi_{iII}^A(r_{12}) = \frac{e_i}{\epsilon_1} \int_0^\infty \left\{ C_{h_d}^A(k) \frac{\exp(-kz_2)}{k} \right\} J_0(ks_{12})kdk \tag{3b}$$

$$\Psi_{iII}^D(r_{12}) = \frac{e_i}{\epsilon_1} \int_0^\infty \left\{ \frac{\exp(-k|z_1 - z_2|)}{k} + \frac{A_{h_d}^D(k)}{k} \exp(-kz_2) \right\}$$

$$\times kJ_0(ks_{12})\, dk \tag{3c}$$

$$\Psi_{iI}^D(r_{12}) = \frac{e_i}{\epsilon_0} \int_0^\infty \left\{ B_{h_d}^A(k) \frac{\exp(kz_2)}{k} + C_0^D(k) \frac{\exp(-kz_2)}{k} \right\} kJ_0(ks_{12})\, dk \tag{3d}$$

The boundary conditions corresponding to the model of the double layer are

$$\Psi^A_{i1}(r_{12}) = \Psi^A_{i\text{II}}(r_{12}) \qquad (z_2 = h_d) \tag{4a}$$

$$\epsilon_0 \left(\frac{d\Psi^A_{i1}(r_{12})}{dz_2} \right)_{z_2=h_d} = \epsilon_1 \left(\frac{d\Psi^A_{i\text{II}}}{dz_2} \right)_{z_2=h_d} \tag{4b}$$

$$\Psi^D_{i1}(r_{12}) = \Psi^D_{i\text{II}}(r_{12}) \qquad (z_2 = h_d) \tag{4c}$$

$$\epsilon_0 \left(\frac{d\Psi^D_{i1}(r_{12})}{dz_2} \right)_{z_2=h_d} = \epsilon_1 \left(\frac{d\Psi^D_{i\text{II}}}{dz_2} \right)_{z_2=h_d} \tag{4d}$$

$$\Psi^A_{i1}(r_{12}) = \Psi^D_{i1}(r_{12}) = 0 \qquad \text{at } z_2 = 0 \tag{5}$$

The system of equations (3) and the conditions (4) and (5) yield the constants $A_0{}^A$, $B^A_{h_d}$, $C^A_{h_d}$, $A^D_{h_d}$, $B^D_{h_d}$, $C_0{}^d$. We refer the reader for the complete solution of the problem to Buff and Stillinger's work [4]. From the results of these authors it may be inferred that

$$\Psi^D_{i\text{II}}(r_{12}) = \frac{e_i}{\epsilon_1} \gamma^D(r_{12}) \tag{6a}$$

where

$$\gamma^D(r_{12}) = \int_0^\infty kJ_0(ks_{12}) \left\{ \frac{\exp(-k\,|\,z_1 - z_2\,|) - \exp[-k(z_1 + z_2)]}{k} \right.$$
$$\left. + \frac{2(\sinh 2h_dk)\exp(-k(z_1 + z_2))]}{k[1 - \delta \exp(-2h_dk)]} \right\} dk \tag{6b}$$

$$\Psi^A_{i1}(s_{12}) = \frac{e_i}{\epsilon_0} \gamma^A(s_{12}) \tag{7a}$$

where

$$\gamma^A(s_{12}) = \int_0^\infty kJ_0(ks_{12}) \left\{ \frac{1 - \exp(-k2z_a)}{k} - \frac{4\delta \exp(-2h_dk)(\sinh zak)^2}{k[1 - \delta \exp(-2h_dk)]} \right\} dk \tag{7b}$$

In the expression (7b) it has been assumed that all charges in Region I are restricted to lie with their center on the IHP. Hence, the location of 1 and 2 is confined to be on the same plane $z = z_a$. Furthermore, we have set $\delta = (\epsilon_1 - \epsilon_0)/(\epsilon_1 + \epsilon_0)$.

A much more tedious method for solving the electrostatic problem would have been to use the images and multiple reflection point of view. Such kinds of solutions can be recovered by expanding the second term of the intergrand in (6b) and (7b) and by performing the integration. Hence, as stated by Buff and Stillinger [4], the closed solution of (6) and (7) constitutes a much more convenient way of summing images of all orders. Yet considering the integrand of (6) and (7), it appears precisely that the first term under the integral accounts for the direct effect of the charge e_i at 1 and for the first-order reflection into the metal. The second term expresses the effect occurring through infinite reflection in the metal and in the h_d boundary.

In solving the system of equations (3c and d) and introducing the condition $z_1 > h_d$, hence considering an ion i far from the interface, the value of ψ_{iII}^D may be shown to be approximately

$$\Psi_{iII}^D(r_{12}) = \frac{e_1}{\epsilon_1} \left\{ \frac{1}{[s_{12}^2 + (z_2' - z_1')^2]^{1/2}} - \frac{1}{[s_{12}^2 + (z_1' + z_2')^2]^{1/2}} \right\} \qquad (8)$$

where

$$z' = z + h_d[(\epsilon_1 - \epsilon_0)/(\epsilon_0] = z + z_v. \qquad (9)$$

For the approximate determination of ψ_{iI}^D, under the conditions $z_1 > h_d$ and $z_2 = z_a$, it is possible to obtain in the same way, by application of the boundary values (4c and d), that

$$\Psi_{i,I}^D(r_{12}) = \frac{e_1}{\epsilon_0} \left\{ \frac{1}{[s_{12}^2 + (z_1' - z_2)^2]^{1/2}} - \frac{1}{[s_{12}^2 + (z_1' + z_2)^2]^{1/2}} \right\} \qquad (10)$$

Let us make clear the meaning of equations (8) and (10). These expressions contain formally the coulomb effect of a point charge and of its single image in the metallic phase. Thus, it appears that for any ion i sufficiently far from the electrode surface ($z_1 > h_d$), the potential $\psi_{iII}^D(r_{12})$ may be obtained in shifting the electrode surface from zero to $z = -h_d[(\epsilon_1 - \epsilon_0)/\epsilon_0]$. For this new location of the origin of z, the potential at any point 2 in Region II is approximately equal to the potential due to a point charge e_i at 1 in a dielectric medium of constant permittivity ϵ_1 for all $z > 0$. From (10), it is inferred that the charge located at 2 (on the IHP) in a medium of dielectric constant ϵ_0 sees the charge i located at 1 in Region II as if the latter charge were in a dielectric of same permittivity ϵ_0 but at a distance increased by z_v. Such virtual effects are analogous to those observed when looking at an object through a glass plate. As mentioned by Buff and Stillinger [4], this kind of virtual effect is not restricted to a model of sharp permittivity steps as illustrated in Fig. 1. For a continuous varying permittivity $\epsilon(z)$, approaching its bulk value $\epsilon(\infty)$

for increasing z, the apparent position of the electrode surface in (8) or of the charge i in (10) is shifted by the distance $|z_v|$, obtained as a solution of the equation

$$0 = \int_{\infty}^{z_v} \frac{dz}{\epsilon(z)} + \int_{z_v}^{\infty} \left[\frac{1}{\epsilon(z)} - \frac{1}{\epsilon(\infty)} \right] dz$$

1.4. Determination of the Average Electrostatic Potential in the Double Layer

Owing to the planar symmetry of the model of the electrochemical double layer considered in Fig. 1, it may be assumed that the average electrostatic potential $\bar{\psi}(r)$, which is the solution of the Poisson equation at r, is only a function of the normal distance z; thus planes parallel to the electrode surface are equipotential in $\bar{\psi}(z)$ and densities of charges corresponding to the adsorbed species are uniformly spread out on planes $z = $ const for $z \geqslant h_d$ and $z = z_a$. The value of $\bar{\psi}(z)$ is readily found by summing the contributions of all charges distributed at the interphase, taking into account the boundary conditions at the metal surface and in the bulk of the solution.

It is convenient to consider first the average potential $\bar{\psi}(z)$ between 0 and h_d. Therefore, one should bear in mind that the point charge potentials $\psi(r)$ (6) and (7) have been calculated by keeping the metal surface ($z = 0$) at zero potential (5). Since the desired value for the average potential $\bar{\psi}(z)$ should correspond to a system where the electrode potential may take any arbitrary value, it is convenient to refer all average electrostatic potentials to their value in the bulk of the solution. Hence,

$$\bar{\psi}(z_a) = A + \sum_{\beta=1}^{\nu} \int_{\mathrm{II}} \psi_{\mathrm{I}}^{D}(r_{ij}) \, \rho_{\beta}^{(1)}(z) \, d\mathbf{r}_i + \rho_{\alpha}^{A} \frac{e_{\alpha}}{\epsilon_0} \int_{z_a} \gamma^A(r_{ij}) \, d\mathbf{s}, \qquad (11)$$

In relation (11) ν is the total number of adsorbed species in the diffuse layer, but only α is specifically adsorbed on z_a; $\rho_{\beta}^{(1)}(r)$ is the singlet probability density (molecules per cubic centimeter) to find a particle of species β at a distance z; ρ_{α}^{A} is the probability density (molecules per square centimeters) to find a particle of species α and charge e_{α} specifically adsorbed at $z = z_a$; and A accounts for the fact that $\bar{\psi}(z)$ must satisfy the boundary conditions

$$\left[\epsilon(z) \frac{\partial \bar{\psi}(z)}{\partial z} \right]_{z \to \infty} = 0 \quad \text{and} \quad \bar{\psi}(z)_{z \to \infty} \to 0 \qquad (12)$$

The solution of (11) is trivial. Elementary electrostatics yields

$$\bar{\psi}(h_d) = \bar{\psi}(0) - \frac{4\pi}{\epsilon_0}(q_a + q^m)h_d + \frac{4\pi}{\epsilon_0}q_a z_a \tag{13}$$

where q_a represents the density of charges specifically adsorbed, $q_a = e_\alpha \rho_\alpha{}^A$, and q^m is the charge density induced by the double layer on the electrode. From the electroneutrality condition it results that $q_a + q^m$ is equal to the charge density induced by the diffuse layer on the electrode surface. Taking into account the assumption made previously of the uniform distribution of charges on the planes $z = $ const in Region II, the application of electrostatics leads to the value of the potential $\bar{\psi}(z)$ at $z = h_d$:

$$\bar{\psi}(h_d) = \frac{4\pi}{\epsilon_1}\sum_{\beta=1}^{\nu}\int_{z_\lambda}^{\infty}(z_d - z)e_\beta \rho_\beta^{(1)}(z)\,dz + 4\pi \frac{(q_a + q^m)(z_d - h_d)}{\epsilon_1} \tag{14}$$

Defining the average field in the inner layer for $h_d \geqslant z \geqslant z_{\shortmid\shortmid}$

$$\bar{E}_I = \frac{4\pi}{\epsilon_0}(q^m + q_a) \tag{15}$$

the average electrostatic potential at $z_{\shortmid\shortmid}$ can now be expressed as

$$\bar{\psi}(z_a) = \bar{E}_I(h_d - z_a) + \frac{4\pi}{\epsilon_1}(q_a + q^m)(z_d - h_d)$$

$$+ \frac{4\pi}{\epsilon_1}\sum_{\beta=1}^{\nu}\int_{z_d}^{\infty}(z_d - z)e_\beta \rho_\beta^{(1)}(z)\,dz \tag{16}$$

The integral appearing in (16) will be evaluated later, in connection with the section concerning the diffuse layer. At high ionic strength ($=10^{-3}$ ml/cm³), the extension of the diffuse layer is small. According to the theory of Krylov and Levich [5], the potential drop in the diffuse layer is substantially steeper than claimed by the Gouy–Chapman model, where the effective thickness of the diffuse region is of the order of $1/\kappa_3$, where κ_3 is the reciprocal Debye length. Hence,

$$\sum_{\beta=1}^{\nu} e_\beta \rho_\beta^{(1)}(z)$$

in (16) may be substituted by $-(q_a + q^m)\,\delta(z - z_d)$, where δ is a Dirac delta function. Consequently, the value of the integral in (16) is zero, which corresponds to the zeroth order approximation in which the entire charge of the diffuse layer is uniformly distributed over the OHP [6].

The average electrostatic potential $\bar{\psi}(z')$ in region $(z' > z_d)$

$$\bar{\psi}(z') = \sum_{\beta=1}^{\infty} \frac{4\pi}{\epsilon_1} e_\beta \int_{z'}^{\infty} (z' - z)\rho_\beta^{(1)}(r)\, dr \qquad (17)$$

is difficult to evaluate and this matter will be discussed in the following sections.

2. THE LOCAL THERMODYNAMIC FORMULATION FOR INHOMOGENEOUS ELECTROCHEMICAL SYSTEMS

2.1. The Local Balance Theory of Prigogine, Mazur, and Defay [7]

The study of interfacial layers in inhomogeneous systems requires the definition of local thermodynamic functions (energy, entropy, free energy). Prigogine and Defay have developed a local balance theory for thermodynamic of polarized systems under the influence of an external electromagnetic field. The general form for local balance of any function F, whose density is $f = F/V$, is given by the fundamental equation

$$\frac{\partial f}{\partial t} = -\operatorname{div}\phi(F) + \sigma(F) \qquad (1)$$

where $\phi(F)$ is the flux associated with f and $\sigma(F)$ is the source that is zero for so-called conserving functions (matter, total energy). Using equations of the type (1) and Maxwell's relation for the electromagnetic field,

$$\mathbf{E} = -\operatorname{grad}\bar{\psi} \qquad \text{etc...} \qquad (2)$$

(where $\mathbf{E}$ is the macroscopic electric field and $\bar{\psi}$ the macroscopic (average) electric potential), those authors obtained a local expression for the density of free energy f of a polarized system (in absence of gravitational field) when it is assumed that all dipoles are in equilibrium with the field. Hence, $f = f(T, \mathbf{P}, \rho_\gamma)$ and

$$df = -s\,dT + \mathbf{E}\,d\mathbf{P} + \sum_{\gamma} \mu_\gamma\, d\rho_\gamma^{(1)} \qquad (3)$$

where s is the density of entropy; $\rho_\gamma^{(1)}$ is the local concentration* of species γ; μ_γ is the chemical potential of γ in the field; $\mathbf{P}$ is the polarization per unit volume of the system and is assumed to be proportional to the field

$$\mathbf{P} = k\mathbf{E} \tag{4}$$

where $k = (\epsilon - 1)/4\pi$ and

$$\epsilon = \epsilon(T, E, \rho_1^{(1)}, ..., \rho_\nu^{(1)})$$

The local permittivity is a function of temperature, of the field and concentrations of all species present in the system. From (3), the following partial derivatives are obtained:

$$\left(\frac{\partial f}{\partial \mathbf{P}}\right)_{T, \rho_\gamma^{(1)}} = \mathbf{E} \tag{5}$$

$$\left(\frac{\partial f}{\partial \rho_\gamma^{(1)}}\right)_{T, \mathbf{P}, \rho_{\gamma \neq \nu}^{(1)}} = \mu_\gamma \tag{6}$$

$$\left(\frac{\partial f}{\partial T}\right)_{\mathbf{P}, \rho_\gamma^{(1)}} = -s \tag{7}$$

In the same way, (6) constitutes the definition of the chemical potential, and integration of (5) leads to

$$f(T, \mathbf{P}, \rho_1^{(1)} ... \rho_\nu^{(1)}) = f(T, \mathbf{P} = 0, \rho_1^{(1)} ... \rho_\nu^{(1)}) + \int_{0,\rho}^{\mathbf{P}} \mathbf{E} \, d\mathbf{P} \tag{8}$$

Using definition (6), it is possible to obtain [8] for $\epsilon = \epsilon(T, E, \rho_1^{(1)} \cdots \rho_\nu^{(1)})$,

$$\mu_\gamma = \mu_{\gamma_0}(T, \mathbf{P} = 0, \rho_1^{(1)} ... \rho_\nu^{(1)}) - \frac{1}{8\pi} \int_{0,\rho}^{E^2} \left(\frac{\partial \epsilon}{\partial \rho_\gamma^{(1)}}\right)_{T(\rho)E} dE^2 \tag{9}$$

The index ρ at the lower boundary of the integral means that the integration has to be performed by maintaining the set of $\rho^{(1)}$ constant. The brackets for (ρ) indicate that all $\rho_1^{(1)} \cdots \rho_\nu^{(1)}$ except $\rho_\gamma^{(1)}$ are constant.†

---

* For convenience $\rho^{(1)}$ is expressed in moles per cubic centimeter in this section only.

† In some recent work [9], a formal expression for free energy and chemical potential was obtained by using instead of isotropic pressure a tensor of viscosity and of capillary forces. The results made it possible to express f and μ_γ in an adsorbed layer as a function of the surface tension and deformation rate of the surface. Unfortunately, these expressions are still rather involved and need some improvement to be of practical use.

The expression of the density of the Gibbs free energy g in a polarized system has been derived by Prigogine et al. [7]. One gets

$$g = g(T, p, \mathbf{P}, \rho_v^{(1)}) = \sum_\gamma \rho_\gamma^{(1)} \mu_\gamma = f + p - \mathbf{EP} \tag{10}$$

in which the meaning of p, the local pressure, will be discussed later. A modification of any variable (at constant p and T) will correspond to a given change δ of (10). By application of equation (3), it appears that the local expression of Gibbs-Duhem's relation in a polarized system is of the form

$$\sum_\gamma \rho_\gamma^{(1)} \delta\mu_\gamma + \mathbf{P}\,\delta\mathbf{E} = 0 \tag{11}$$

2.2. The Definition of Pressure at the Interphase [10, 11]

The chemical potential at zero electric field for any solute s in dilute solution of small compressibility ($\rho_s^{(1)} \ll \rho_1^{(1)}$, where the subscript 1 indicates solvent) is given by

$$\mu_{s0} = \eta_{s0}(T) + v_{s0}^* p_0 - RT \ln \rho_{10}^{(1)}(T, p_0) + RT \ln f_{s0}\rho_s^{(1)} \tag{12}$$

where f_{s0} is the local activity coefficient related to the molal fraction of nonhydrated molecules at zero polarization and at pressure p_0; $\rho_{10}^{(1)}$ is the concentration of the solvent under the same conditions; and v_{s0} is the standard specific molar volume. The pressure p_0 or Helmholtz pressure is defined at zero polarization but for the same local conditions of temperature and concentration as those existing under the effect of the field. This pressure p_0 is obviously different from the pressure p appearing in the condition of hydrodynamic equilibrium. If the electric acting force is chosen to be

$$q\mathbf{E} + \text{grad } \mathbf{E} \cdot \mathbf{P} \tag{13}$$

where q is the density of electric charge, at hydrodynamic equilibrium we obtain at a plane interface, normal to the axis z [10,12],

$$\frac{\partial p}{\partial z} = qE_z + P_z \frac{\partial E_z}{\partial z} \tag{14}$$

Using the Poisson equation,

$$\text{div } \mathbf{D} = \left[\epsilon \frac{\partial E}{\partial z} + E_z \frac{\partial \epsilon}{\partial z} \right] = 4\pi q \tag{15}$$

it is easy to perform the integration of (14) bounded by $p = p'$ at $E = 0$ in the bulk of the solution and by any point at the interphase where $E \neq 0$.

$$p - p' = \frac{E^2}{8\pi} (2\epsilon - 1) \qquad (16)$$

The pressure p is a scalar quantity. Yet we must keep in mind that although a polarized fluid is isotropic with respect to the chemical and electrical properties as a whole, it behaves with respect to one of these properties alone as anisotropic with symmetry about the polarization direction [13]. Accordingly, it is known from electrostatics that the pressure p_z across a surface normal to the polarization vector exceeds the local isotropic hydrodynamic pressure p by $2\pi P^2$. Thus, it is found [12] on account of (16) that

$$p_z = p' + \epsilon^2 E^2/8\pi \qquad (17)$$

If $\epsilon = 6$ and $E = 10^7$ V/cm, $(p_z - p') = 10^3$ atm.

It is sometimes emphasized that the electrostatic pressure exerted by the remaining double layer on the adsorbed dipole film may influence the value of the local polarizability in this layer. This effect may be readily evaluated at h_d owing to the fact that the polarization in the region $z > h_d$ is proportional to the field. Hence, for the electric field $(4\pi/\epsilon_1)(q_a + q^m) = E(h_d)$, the relation (17) leads to

$$p_z = p' + 18\pi(n_{\mu_q})^2 \qquad \text{[bars]} \qquad (18)$$

where n_{μ_q} is expressed in μCb/cm^2. The difference $p_z - p'$ so computed is of the order of 10^3 atm. According to the high incompressibility of water, the change of permittivity under these pressures may be still small. Owen et al. [14] have obtained the ratio

$$\frac{\epsilon_1(20°C, 980.7 \text{ bars})}{\epsilon_1(20°C, 1 \text{ bar})} = 1.09392$$

for bulk water.

The difference of pressure $p_0 - p$ may be also obtained and was shown to be

$$p_0 - p = \frac{1}{8\pi} \int_{0,\rho}^{E^2} \left[\sum_\gamma \rho_\gamma^{(1)} \left(\frac{\partial \epsilon}{\partial \rho_\gamma^{(1)}} \right)_{T(_0)E} - (\epsilon - 1) \right] dE^2 \qquad (19)$$

The physical meaning of (19) may be understood easily for a one-component system. Using Kirkwood's model [15] for the permittivity ϵ,

m being the permanent dipole moment of the molecules in the medium and M the electric moment for a macroscopic sphere including an average configuration of particles around a central fixed particle, we have

$$p_0 = p - \frac{\epsilon E^2}{8\pi} \left(\frac{\partial \ln mM}{\partial \ln \rho^{(1)}} \right)_T \tag{20}$$

Since $(\partial \ln mM/\partial \ln \rho^{(1)})_T$ is different from zero only for statistically dependent molecules, the difference of pressure $p_0 - p$ is related to the work done by modifying the correlations between molecules submitted to an external field E. In case of water, a rough evaluation of $(\partial \ln nM/\partial \ln \rho^{(1)})_T$ using the data of Owen, Miller, Milner, and Cogan [14] for $(\partial \ln \epsilon/\delta p_0)$ shows that $(\partial \ln mM/\delta \ln \rho^{(1)})_{T=25^\circ C} \simeq 0.03$.

2.3. The Definition of the Electrochemical Potential

The local balance of entropy in any system may be shown to be of the form

$$\frac{\partial s}{\partial t} = - \operatorname{div} \phi(s) + \sum_i \left(\frac{A_i v^i}{T} \right) \tag{21}$$

Where A_i is called the affinity of any process i and v^i is the rate. The fundamental equation of production of entropy (or source of entropy) is

$$T\sigma(s) = \sum_i A_i v^i > 0 \text{ (nonstationary process)}$$

$$= 0 \qquad \text{(stationary or equilibrium process)} \tag{22}$$

Prigogine, Mazur, and Defay [7] have shown that in the case of pure chemical reactions,

$$A_c = - \sum_\gamma v^\gamma \mu_\gamma \tag{23}$$

where v^γ is the stoichiometric coefficient of component γ. In the same way they obtained for the affinity of diffusion in the absence of magnetic influence:

$$A_D = - \sum_\gamma \operatorname{grad} (e_\gamma \psi + \mu_\gamma) = - \sum_\gamma \operatorname{grad} \tilde{\mu}_\gamma \tag{24}$$

where $\tilde{\mu}_\gamma$ is now the electrochemical potential and e_γ is the electrical charge. The condition of hydrodynamic (or diffusion) equilibrium (14) is satisfied if

$$\operatorname{grad} \tilde{\mu}_1 = \operatorname{grad} \tilde{\mu}_2 = \cdots \operatorname{grad} \tilde{\mu}_x = 0 \tag{25}$$

as it may be easily proved by introducing (24) into (11) and comparing with (14) at constant p. Thus, at diffusion equilibrium, the following relation is obtained:

$$\mu'_\gamma - \mu''_\gamma = e_\gamma(\bar{\psi}'' - \bar{\psi}') \tag{26}$$

Consequently, the long-range effects appearing in the right-hand term of (26) are balanced by the difference of local physicochemical state. The influence of the polarization factor in equation (9) is obviously a local effect.

The problem of decomposition of the electrochemical potential into a chemical and electrical term is in some way academic, because both effects will be coupled in most cases (in the presence of magnetic effects the problem is much more complicated). Nevertheless, the distinction between local and long-range coulombic effects as pointed out by Lange [16], as well as the value of the affinity of diffusion obtained through the local balance method, should be used as good arguments for accepting the above stated decomposition [17,18]. To conclude, it may be pointed out that the relations between local concentrations, mean electric potentials, and activities (9), (12), (26) as dictated by extension of thermodynamic reasoning to the "local" domain, are also encountered in a statistical treatment [4] based on Mayer's cluster theory. This supplies one microscopic criterion for applicability of the local thermodynamic balance and for the formulation adopted.

2.4. The Distribution Function at Equilibrium [12]

By inserting equations (9), (12), (16), and (19) into (26), one obtains the equilibrium distribution of solute particles between some point of the diffuse layer and the bulk of the system:

$$\frac{\rho_s^{(1)}(z)}{\rho_s^{(1)'}} = \frac{f'_{s0}}{f_{s0}} \exp\left\{\left(\frac{v^*_{s0}}{RT}\right) - \frac{\epsilon E^2}{8\pi} - \frac{1}{8\pi} \int_{0,\rho}^{E} \left[\sum_\gamma \rho_\gamma^{(1)} \left(\frac{\partial \epsilon}{\partial \rho_\gamma^{(1)}}\right)_{T,E,(\rho)} \right.\right.$$

$$\left.\left. - \frac{1}{v^*_{s0}} \left(\frac{\partial \epsilon}{\partial \rho_s^{(1)}}\right)_{T,E,(\rho)} + \frac{1}{2} E \left(\frac{\partial \epsilon}{\partial E}\right)_{T,\rho} \right] dE^2 - \frac{e_s \bar{\psi}(2)}{kT} \right\} \tag{27}$$

where $\bar{\psi} - \bar{\psi}' = \bar{\psi}(z)$ and e_s is the value of the charge of the solute.

The relation (27) will be used later, coupled to the Poisson equation (15) [12]. We wish here only to call attention to the following consequence of (27). Let us consider the distribution of two substances A, B of different volume and polarizability. It may be shown through (27) that the more polarizable species A with

$$\left(\frac{\partial \epsilon}{\partial \rho_A^{(1)}}\right)_{T,(\rho)} > \left(\frac{\partial \epsilon}{\partial \rho_B}\right)_{T,(\rho)}$$

will be the most concentrated in the region where the potential distribution $\bar{\psi}$ is the most inhomogeneous [19,20]. This effect may be increased if $v_{A,0}^* < v_{B,0}^*$.

3. THEORIES OF THE IONIC DOUBLE LAYER

3.1. General Evolution of the Theoretical Concepts of the Difluse Layer

Gouy [21] and Chapman [22] were the first to introduce the Poisson–Boltzmann equation at the beginning of this century for interpreting the behavior of colloidal systems. Through this relation they tried to improve the old theory of the parallel plate capacitor for the double layer that was given by Helmholtz [23] in 1879. In order to integrate the Poisson–Boltzmann equation, Gouy and Chapman assumed a medium of nonpolarizable charges suspended in a dielectric continuum.

Later, Debye and Hückel [24] solved the Poisson–Boltzmann equation to account for the average electrostatic potential of ionic interaction in a solution of strong electrolytes, but these authors, after having expanded the Boltzmann exponential, limited the integration to the linear terms in the average electrostatic potential. Under these conditions, the solution of Debye and Hückel may be considered one of the most outstanding achievements of modern theoretical chemistry since it provided a general understanding of the typical features of strong electrolytes, arising from the long-range coulombic forces between the ions.

The success of the strong electrolyte theory gave rise rapidly to some questions about the precise theoretical foundations of the Poisson–Boltzmann equation. First of all, however, the conditions of internal consistency of a Debye–Hückel type of solution were demonstrated [25,26]. Hence, it was readily shown that only the retention of linear terms in the average electrostatic potential, as done in the treatment of Debye–Hückel, leads to an agreement with (1) the condition of integrability that must be satisfied by the electrostatic part of the excess free energy and (2) the principle of symmetry of the pairwise potential of mean force upon which Debye and Hückel's assumption is initially based.

Nevertheless, as pointed out later by Kirkwood and Poirier [27], the above-mentioned analysis did not prove in fact that the relation of Debye and Hückel was valid and its basis deducible from a real statistical mechanical argument. To reexamine the fundamentals of this very important problem, two major theoretical directions were adopted. The cluster theory of Mayer constitutes one of these useful approaches [28]. The other way has been tackled by Kirkwood and Poirier [27] by means of developing the potential of average force in power series of an ionic charging parameter. Both methods were introduced very recently in the theory of the electrochemical interphase and have given a better self-consistent, although limited, solution for the charge and potential distribution in the diffuse layer.

Apart from the statistical–mechanical viewpoint so adopted, some useful but nevertheless less accurate and sometimes intuitively appealing approaches have been developed. In most of these works the validity of the Poisson–Boltzmann equation is basically assumed and different corrections are introduced that may account for some plausible effects as the ionic excluded volume [29–34,38], the dielectric saturation [26,34–36,38,89], and the polarizability of the particles and their degree of hydration [37,38]. More recently, a rather general understanding for this kind of approach has been given on the basis of the local thermodynamic theory [12] as presented in Section 2.

The determination of the molecular structure of the diffuse part of the double layer appears essential for understanding a great number of electrochemical phenomena. As shown by Grahame [39], this knowledge is generally indispensable if one attempts to recover the configuration of the inner part of the double layer from interfacial capacity measurements or from electrocapillary curves. The potential distribution in the diffuse layer is also introduced in the works of Frumkin et al. [40] and of Gierst [41] for calculating the local amount of reacting species at their closest approach to the electrode. Furthermore, the value of the field in the double layer may affect the chemical constitution (ion pairing etc.) of the adsorbing and discharging species.

3.2. Preliminaries on the Statistical Mechanical Formulation in Fluids [42]

3.2.1 *The Set Notation* [43].

(i)   The composition set:

$$\mathbf{n} \equiv n_1 , n_2 ,..., n_\beta ,..., n_\nu \tag{1}$$

In this set the elements separated by commas are the number of particles of each species 1, 2, β, and ν, where ν is also the total number of species that may be present.

(ii)   The coordinate set:

$$\{\mathbf{n}\} \equiv \{1_1\}, \{2_1\} ,..., \{n_1\} \\ \{1_2\}, \{2_2\} ,..., \{n_2\} \\ \{1_\nu\}, \{2_\nu\} ,..., \{n_\nu\} \tag{2}$$

Each of the elements of these sets specifies the complete set of coordinates of one particle in the vessel. This implies that all particles are distinguishable, even those of the same species.

(iii)   The concentration set:

$$\mathbf{\rho} \equiv \rho_1 , \rho_2 ,..., \rho_\nu \tag{3}$$

This set is closely related to $\mathbf{n}$, the concentration of species s being defined by $\rho_s \equiv n_s/V$, where V is the volume of the vessel.

(iv)   The chemical potential set:

$$\mathbf{\mu} = \mu_1 , \mu_2 ,..., \mu_\nu \tag{4}$$

(v)   Definition of some important operations:

$$n \equiv n_1 + n_2 + \cdots + n_\nu \quad \text{(sum of composition elements)} \tag{5}$$

$$\mathbf{n}! \equiv n_1!! \cdots n_2! \cdots n_\nu! \quad \text{(product of factorials)} \tag{6a}$$

$$[\mathbf{n} - (\alpha, \beta)]! = n_1! \, n_2! \cdots [n_\alpha - 1]! \cdots [n_\beta - 1]! \cdots n_\nu! \tag{6b}$$

$$\mathbf{c}^{\mathbf{n}} \equiv c_1^{n_1} \cdot c_2^{n_2} \cdots c_\nu^{\,n} \quad \text{(product of exponentials)} \tag{7}$$

$$\mathbf{n\mu} \equiv n_1\mu_1 + n_2\mu_2 + \cdots + n_\nu\mu_\nu \quad \text{(Gibbs free energy)} \tag{8}$$

3.2.2. *Definition of Probability Density in the Canonical Ensemble.*
The definition of the probability density $P_s(\{\mathbf{r}\})$ in the coordinate set $\{\mathbf{r} + \mathbf{n}\} \equiv \{\mathbf{m}\}$, for observing the particles of the composition set $\mathbf{n}$

at unspecified coordinates but particles of **r** at particular coordinates $\{\mathbf{r}\}$, is given by the expression

$$P_s(\{\mathbf{r}\}) \, d\{\mathbf{r}\} = d\{\mathbf{r}\} \frac{1}{Z} \int_V \exp\left[-\frac{U(\{\mathbf{m}\})}{kT}\right] d\{\mathbf{n}\} \tag{9}$$

where Z is the configuration integral and $U(\{\mathbf{m}\})$ is the direct potential. This potential may be expanded in terms of direct component potentials

$$U(\{\mathbf{m}\}) = \sum_{\substack{\{\mathbf{t}\} \leqslant \{\mathbf{n}\} \\ t \geqslant 2}} u_t(\{\mathbf{t}\}) \tag{10}$$

Here $u_t(\{\mathbf{t}\})$ is the reversible isothermic work required to bring the particles at position $\{\mathbf{t}\}$ from some state where these particles do not interact. By definition, each element of the set $\{\mathbf{t}\}$ interacts with the other particles of the set but does not interact with the remaining particles $(\mathbf{m} - \mathbf{t})$, which accordingly may be considered as independent. Hence, one may write for a dilute medium by assuming $U(\{\mathbf{m}\}) \simeq u_r(\{\mathbf{r}\})$ and neglecting the interactions of any element of $\{\mathbf{r}\}$ with $\{\mathbf{n}\}$:

$$P_s(\{\mathbf{r}\}) \simeq \frac{V^n}{V^m} \exp\left[-\frac{u_r(\{\mathbf{r}\})}{kT}\right] = \frac{1}{V^r} \exp\left[-\frac{u_r(\{\mathbf{r}\})}{kT}\right] \tag{11}$$

In nondilute medium, it becomes necessary to introduce the potential of mean force $W_r(\{\mathbf{r}\})$, which may be defined through the following relation:

$$P_s(\{\mathbf{r}\}) = \frac{1}{Z} \int \exp\left[-\frac{U(\{\mathbf{m}\})}{kT}\right] d\{\mathbf{n}\} = \frac{1}{V^r} \exp -\frac{W_r(\{\mathbf{r}\})}{kT} \tag{12}$$

It is readily shown that $W_r(\{\mathbf{r}\})$ derives (1) from the forces through which the particles of set $\{\mathbf{r}\}$ are directly acting one on each other and (2) from the statistical averaging of the forces exerted by the remaining set $\{\mathbf{n}\}$ on the particles contained in $\{\mathbf{r}\}$.

In the case of solutions it is useful to define the sets $\{\mathbf{m}\}$, **m**, **e** for the solute only and to include in the direct potential some component of the potential of mean force expressing the interactions between a cluster of solute particles and the solvent. For example, any pairwise potential in an electrolyte solution can be given in the following way:

$$u_{ij}(\{ij\}) = \frac{e_i e_i}{\epsilon r_{ij}} + u_{ij}^*(\{ij\}) \tag{13}$$

r_{ij} being the distance between i and j. The effects of the solvent, considered as a dielectric continuum, are now introduced through the permittivity ϵ. The influence of the molecular structure of the solvent is included in the remaining term u_{ij}^*, which also contains the mutual polarization work of the ions as well as the van der Waals interactions at short distances [effects of $o(r^{-4})$ and less] and the very short-range quantic repulsive effects.

3.2.3. *The Correlation and Radial Distribution Function.* The probability density $P_s(\{r\})$ is specific for each particle defined by one element of the coordinate set $\{r\}$. The corresponding generic probability density or distribution function $\rho^{(r)}(r)$ depends only on the composition set r and is obtained by multiplying $P_s(\{r\})$ by the number of ways of assigning the particles of composition set $(n + r)$ to the coordinates $\{n + r\}$ and dividing by the number of integrals of equation (12) that should be the same even if the particles are distinguishable. Thus, so far

$$\rho^{(r)}(r) \equiv \frac{(r + n)!}{n!} P_s(\{r\}) \tag{14}$$

It becomes now possible to rewrite $\rho^{(r)}(r)$ by introducing the generic correlation function $g^{(r)}(r)$ that plays an important role in the next development:

$$\rho^{(r)}(r) = \rho^{(r)} g^{(r)}(r) = \left(\frac{m}{V}\right)^r g^{(r)}(r) = \frac{(r + n)!}{n! \, V^r} \exp\left[-\frac{W_r(\{r\})}{kT}\right] \tag{15}$$

It is easily shown from the above definition of $g^{(r)}(r)$ that for $r \ll n$ (dilute medium) one obtains $(r + n)!/n! \simeq m^r$; thus,

$$g^{(r)}(r) = \exp - \frac{W(\{r\})}{kT} \tag{16}$$

Furthermore, the pair correlation function $g^{(r)}(r) \equiv g^{(2)}(\alpha, \gamma)$, if two particles $\alpha_{(1)}$ and $\gamma_{(2)}$ are distant by r_{12}, is just the radial distribution function $g^{(r)}(r_{12})$, which is related to the probability of observing a second unspecified particle γ at a distance r_{12} from α. As a matter of fact, according to the definition of $P_s(\{\alpha_{(1)}, \gamma_{(2)}\})$ and (14), it turns out that the probability of finding an unspecified particle γ in dr_{12} at a distance r_{12}

from α, if α is fixed anywhere in the vessel of volume V, is given by

$$d\mathbf{r}_{12} \frac{[\mathbf{m} - \alpha]!}{[\mathbf{m} - [\alpha, \sigma]]!} \int_V d\mathbf{r}_1 P_s(\{\alpha_{(1)}, \sigma_{(2)}\}) = V m_\alpha P_s(\{r_{12}, \alpha, \sigma\}) \, d\mathbf{r}_{12}$$

$$= V \frac{\rho^{(2)}(\mathbf{r}_{12}, \alpha, \sigma) \, d\mathbf{r}_{12}}{m_\alpha} \qquad (17)$$

$$= \rho_\alpha g^{(r)}(\mathbf{r}_{12}, \alpha, \sigma) \, d\mathbf{r}_{12}$$

3.2.4. *Kirkwood's superposition approximation.* Kirkwood has admitted that any decomposition of type (10) for the direct potential may be applied to the potential of average force as well. Thus,

$$W(\{\mathbf{t}\}) = \sum_{\substack{\{\mathbf{t}\} \leqslant \{\mathbf{r}\} \\ t \geqslant 2}} W_t(\{\mathbf{t}\}) \qquad (18)$$

The case in which $r = 3$ is of particular interest. Then we have

$$W(\{ijk\}) = W_{ij} + W_{jk} + W_{ki} + W_{ijk} \qquad (19)$$

In the Kirkwood approximation, the component potential W_{ijk}, also called the superposition defect, is zero. Thus, it is stated that the average force acting on a third particle in the neighborhood of a pair is the sum of the average forces acting on it if each particle of the pair were present alone. It is important to keep in mind that, by definition, the pairwise potential of mean force $W_{ij}(r_{ij})$ must obey the relation

$$W_{ij}(r_{ij}) = W_{ji}(r_{ji}) \qquad (20)$$

3.2.5. *Distribution in the Grand Ensemble.* Up to this point the distribution functions (9), (11), and (15) have been derived for the canonical ensemble. The grand canonical ensemble often appears much more convenient in the treatment of a two-phase system (as it is the case of adsorption from the bulk phase to the interphase) for the reason that the condition of exchange equilibrium between the phases may already be included in the initial distribution function. The expression for the generic distribution function $\rho^{(r)}(\mathbf{r})$ now becomes

$$\rho^{(r)}(\mathbf{r}) = \frac{1}{\Xi} \exp \frac{\mathbf{r}(\mu - \mu^0)}{kT} \sum_n \exp \frac{\mathbf{n}(\mu - \mu^0)}{kT} \frac{1}{\mathbf{n}!} \int \exp - \frac{U(\{\mathbf{m}\}) \, d\{\mathbf{n}\}}{kT}$$

$$= \rho^r g^{(r)}(\mathbf{r}) \qquad (21)$$

in which Ξ is the grand partition function and μ^0 represents the set of standard chemical potential.

3.2.6. *Average Electrostatic Potential and Free Energy.* The Helmholtz free energy of mixing is related to the canonical configurational partition function Z through the fundamental relation

$$F_m = -kT \ln Z \qquad (22)$$

The corresponding expression in the grand ensemble yields

$$\begin{matrix} \text{Bulk phase} & pV \\ \text{Surface phase} & p'\Omega \end{matrix} \Bigg| = kT \ln \Xi \qquad (23)$$

in which p and V are respectively the pressure (or osmotic pressure in the case of a solution) and volume of the bulk phase and p' and Ω are the spreading pressure and area of a surface phase.

The differential change of F_m per increase of the charge e_α of some ion α at position $\{\alpha\}$ is derived from (10), (13), and (22); thus, assuming no short-range interaction dependence on e_α, one has*

$$\left(\frac{\partial F_m}{\partial e_\alpha}\right)_{T,V,\mathbf{n}} = -\frac{kT}{Z}\left[\frac{\partial}{\partial e_\alpha}\int \exp[-U(\{\mathbf{m}\})/kT]\,d\{\mathbf{m}\}\right]_{T,V,\mathbf{n}}$$

$$= \sum_{\{\alpha,\gamma\}}\frac{1}{\epsilon}\int\frac{e_\gamma}{r_{\alpha\gamma}(\alpha,\gamma)}\frac{\exp[-U(\{\mathbf{m}\})/kT]}{Z}\,d\{\mathbf{m} - \alpha,\gamma\}\,d\{\alpha,\gamma\} \qquad (24)$$

where the summation is over all specific pairs obtained by coupling one of the remaining distinct particles with α. We note that

$$\frac{[\mathbf{m} - \alpha]!}{[\mathbf{m} - [\alpha,\gamma]]!} = \begin{cases} m_\gamma & \text{for } \gamma \neq \alpha \\ m_\gamma - 1 & \text{for } \gamma = \alpha \end{cases} \qquad (25)$$

expresses the number of ways to distribute $m - \alpha$ distinguishable particles between the site $\{\gamma\}$ and the remaining sites of the system (without $\{\alpha\}$), so that only some nonspecified particle γ is assigned to be at $\{\gamma\}$. This corresponds also to the number of partitions giving the same result on integration of (24). Hence, we are now in position to substitute in (24) the sum over particles by a sum over the number of species. Thus, taking into account (12), one obtains

$$\left(\frac{\partial F_m}{\partial e_\alpha}\right)_{T,V} = \sum_\gamma \frac{(\mathbf{m}-\alpha)!}{[\mathbf{m}-(\alpha,\gamma)]!}\frac{1}{\epsilon V^2}\int\frac{e_\gamma}{r_{\alpha\gamma}(\alpha,\gamma)}\exp[-W_{\alpha\gamma}(\{\alpha,\gamma\})/kT]d\{\alpha,\gamma\}$$

$$= \sum_\gamma \int \frac{e_\gamma}{\epsilon r_{\alpha\gamma}(\alpha,\gamma)}\rho_\gamma \exp[-W_{\alpha\gamma}(\{\alpha,\gamma\})/kT]\,d\mathbf{r}_{\alpha\gamma} = \bar{\psi}_\alpha \qquad (26)$$

---

* This does not mean that the integrability property of F_m depends on a given model of interaction. As a matter of fact, this property is far more general.

The meaning of (26) follows from considerations given in relation with (17). Thus, the integrand of (26) expresses a coulombic effect, at the place occupied by α, exerted by the probable charge density of γ at $\{\gamma\}$. After integration and summation over all species in the system, one obtains the electrostatic average potential at the place of α. A similar relation to (26) can be derived for all ionic species of the system. That is to say that the total differential of F_m yields

$$(dF_m)_{T,V,\mathbf{n}} = \overline{\psi_\alpha}\, de_\alpha + \overline{\psi_\beta}\, de_\beta + \cdots \tag{27}$$

The free energy is a single-valued function of the equilibrium of the system. Therefore, (27) indicates that dF_m is an exact differential for which the curl condition is

$$\frac{\partial \overline{\psi_\alpha}}{\partial e_\beta} = \frac{\partial \overline{\psi_\beta}}{\partial e_\alpha} = \cdots \tag{28}$$

The integrability property (28) leads to an important consequence. It is that the path chosen for the integration of (26), starting from an initial state without interaction, is fully arbitrary. In other words, whatever may be the charging process, the same final result must be obtained [26] for the excess electrostatic free energy.

3.3. The Differential Equation Method of Debye and Hückel

The average electrostatic potential at any point in the solution would vanish unless one refers to a specific ion and looks from its relative position to the remaining assembly. Let α be the reference ion that is considered to be maintained at a given position $\{1\}$. The average electrostatic potential at $\{2\}$ results thus from the effect of the average distribution of the remaining ions determined by the presence of α at 1 and from the direct coulombic contribution of the particle α. This leads to the Debye potential $\bar{\psi}_D^\alpha(r_{12})$; thus, so far

$$\bar{\psi}_D^\alpha(r_{12}) = \sum_\gamma \int \frac{e_\gamma}{\epsilon r_{23}} \rho_\gamma g^{(2)}(r_{13}, \alpha, \gamma)\, d\mathbf{r}_{13} + \frac{e_\alpha}{\epsilon r_{12}} \tag{29}$$

With respect to the electrostatic free excess energy evaluation, it is important to note that the first right-hand side term of (29) yields back (26) under the condition that $r_{12} \to 0$.

Upon introduction of the operator ∇^2, (29) transforms into

$$\nabla^2 \bar{\psi}_D^{\alpha}(r_{12}) = -\frac{4\pi}{\epsilon} \sum_{\gamma \neq \alpha} e_\gamma \rho_\gamma g^{(2)}(r_{12}, \alpha, \gamma) - \frac{4\pi e_\alpha}{\epsilon} \delta(r_{12})$$

$$= -\frac{4\pi}{\epsilon} [(q(r_{12}) + e_\alpha \delta(r_{12}))] \tag{30}$$

where $\delta(r_{12})$ is the Dirac delta function in three dimensions. It was the brilliant assumption of Debye to insert (29) into the Poisson equation and to deduce the density of charge $q(r_{12})$ at a distance r_{12} from the selected ion α by introducing the statistical expression for the radial distribution function (17). The basic approximation of Debye and Hückel then lies in the fact that the pairwise potential of average force $W_{12}(\{\alpha, \gamma\})$ is replaced by $e_\gamma \bar{\psi}_D^{\alpha}(r_{12})$; thus,

$$q(r_{12}) = \sum_{\gamma \neq \alpha} e_\gamma \rho_\gamma eap - \frac{e_\gamma \bar{\psi}_D^{\alpha}(r_{12})}{kT} \tag{31}$$

The serious effect of this approximation has been often discussed [26] and was pointed out in Section (3.1). A fast scrutiny of this problem enables us to state the two following points: (1) The assumption of Debye and Hückel corresponds to the choice of a general model of interactions based on the perfect superposition of pure coulombic effects acting at 2. The property of superposition lies in the strict definition of the average electrostatic potential as given in equation (29). Such a superposition would be quite valid if the density of ions, which are exhibiting a sphere of exclusion of mean radius a, is not too near the close packed density, i.e.,

$$I_1 \equiv \frac{a^3}{\sqrt{2}} \sum_\gamma \rho_\gamma \ll 1 \tag{32}$$

(2) The linearization condition: Whereas the electrostatic law (30) is based on the principle of superposition of charge densities, which means that doubling $\bar{\psi}_D^{\alpha}(r_{12})$ corresponds to double the charge $q(r_{12})$, the same proportionality obviously does not hold for the microscopic function of distribution (31) *in which the approximation of Debye has been assumed*; hence, doubling the exponent of (31) would not give a double value of $q(r_{12})$, unless (31) is expanded and the linear term is retained. Such a linearization of the Boltzmann term is possible whenever

$$I_2 = \frac{e_\gamma \bar{\psi}_D^{\alpha}(r_{12})}{kT} \ll 1 \tag{33}$$

The condition (33) must hold over the entire range of integration of (30) and (31); hence, it may be stated that, according to the size parameter, a, one has, in principle, at the lower limit of integration, separately the following inequalities for each term of (29):

$$I_3 = \left| \frac{e_\gamma e_\alpha}{kT\epsilon a} \right| \ll 1 \tag{34}$$

$$I_4 = \lim_{r_{12} \to a} \frac{1}{kT} \left(\left| e_\gamma \bar{\psi}_D{}^\alpha(r_{12}) - \frac{e_\gamma e_\alpha}{\epsilon r_{12}} \right| \right) \ll 1 \tag{35a}$$

The discrepancy between macroscopic and microscopic behavior in the case where no linearization can be assumed is best reflected through the fact that the final result neither conforms to the basic condition (20) nor (28).

The linear solution of the Poisson–Boltzmann equation (30) and (31) is well known. One has at sufficiently high dilution (neglecting the size parameter a),

$$\bar{\psi}_D{}^\alpha(r_{12}) = \frac{e_\alpha}{\epsilon r_{12}} \exp\left[-\kappa_3 r_{12}\right] \tag{36}$$

where κ_3 is the reciprocal distance of Debye and Hückel,

$$\kappa_3{}^2 \equiv \frac{4\pi}{\epsilon kT} \sum_\gamma \rho_\gamma e_\gamma{}^2 \tag{37}$$

Taking into account (36), the inequality (35) can be written

$$I_4 \equiv \left| \frac{\kappa_3 e_\alpha e_\gamma}{kT\epsilon} \right| \ll 1 \tag{35b}$$

It appears from the previous discussion that any extension of the theory of strong electrolytes is best accomplished by avoiding the most questionable *a priori* postulate (31). This viewpoint characterized the methods of Mayer [28] and Kirkwood and Poirier [27]. Their approach provides at the same time a deeper insight, on statistical mechanical grounds, into the fundamentals of Debye's assumption.

3.4. The Gouy Potential in the Diffuse Layer

Having outlined the general ideas and problems in the case of ionic solution, we now proceed to apply similar concepts to the double-layer region. The systems considered in the following section contain generally a symmetric 1-1-electrolyte, and the electrode is always treated as an infinite plane.

3.4.1. *The Integrability Condition in the Case of an Inhomogeneous System.* Under the influence of an external field the distribution of particles becomes inhomogeneous. Thus, it appears necessary to introduce a local thermodynamic formulation (cf. Section 2) and to determine in this way the state of the system at a given place $v(r)$ that is very small but still of macroscopic dimension. The thermodynamic functions so defined do not depend explicitly on variables characterizing other parts of the system. For example, the differential of the local electrochemical free energy of Gibbs $\tilde{g}[v(r)]$ at constant local composition (and constant p, **P**, and T) may be written in the adsorbed phase:

$$\{\delta \tilde{g}[v(\mathbf{r})]\}_{\rho^{(1)}, T, p} = \sum_{\gamma} \{\rho_{\gamma}^{(1)}[v(\mathbf{r})] - \rho_{\gamma}'\}\{\delta \tilde{\mu}_{\gamma}[v(\mathbf{r})]\}_{\rho^{(1)}, T, \mathbf{P}, p}$$

$$= \sum_{\gamma} \{\rho_{\gamma}^{(1)}[v(\mathbf{r})] - \rho_{\gamma}'\} \left. \frac{\partial \tilde{\mu}_{\gamma}[v(\mathbf{r})]}{\partial e_{\gamma}} \right|_{\rho^{(1)}, T, \mathbf{P}, p} \delta e_{\gamma} \qquad (38)^*$$

where ρ_{γ}' is the bulk concentration and $\rho_{\gamma}^{(1)}[v(r)]$ is the singlet distribution function of species γ at $v(r)$.

The statistical mechanical expression for the electrochemical potential $\tilde{\mu}$ in an inhomogeneous system has been derived by Buff and Stillinger [4] starting from the expression of the singlet function $\rho_{\gamma}^{(1)}$ in the grand canonical ensemble [equation (21)]; these authors obtained the following relation for $\tilde{\mu}$ of α at (1) [situated into $v(\mathbf{r})$]:

$$\tilde{\mu}_{\alpha}(1) = \mu_{\alpha}{}^0(1) + e_{\alpha}\psi_{(1)}^{\text{ext}}$$

$$- kT \sum_{(n-\alpha) \geqslant 1} \frac{1}{[\mathbf{n} - \alpha]!} \int S\{\mathbf{n}\}\rho_2^{(1)}(\{2\})$$

$$\times \rho_3^{(1)}(\{3\}) \cdots \rho_{n_\nu}^{(1)}(\{n_\nu\}) \, d\{\mathbf{n} - \alpha\} + kT \ln \rho_{\alpha}^{(1)} \qquad (39)$$

where $S\{\mathbf{n}\}$ represents a cluster sum for at least double-connected graphs built on a skeleton of **n** distinguishable particles at position $\{1\}, \{2\},..., \{n_\nu\}$, which are, respectively, 1, 2,..., n_ν.[†] Furthermore, the coordinate set $\{\mathbf{n} - \alpha\}$ is integrated over the entire space, this means

---

* The partial differentiation in (38) may be formally considered although the sets of variables **e** and $\mathbf{\rho}^{(1)}$ are physically dependent [cf. also equation (47)]. Thus e_{γ} is used here as an electrical charging parameter of particle γ ($e_{\gamma}\xi_{\gamma}$ in Section 3.4.4), at constant composition.

† For the terminology and techniques of cluster theory, compare Friedman [43].

that all the configurations remain rooted at the point (1) where α is located; $\psi_{(1)}^{\text{ext}}$ is the coulomb potential at (1), arising directly from the effect of the external electric field.

In order to proceed further, it is convenient to rewrite the last right-hand-side term of (39) in such a way that in turn differentiation by the bond between α at (1) and one remaining particle at (2) is performed in each graph. Thus, one gets on account of (25),

$$
kT \sum_{\gamma=1} \sum_{(\mathbf{n}-\alpha,\gamma)\geqslant 0} \frac{1}{[\mathbf{n}-\alpha]!} \int d\mathbf{r}_{12}\, \rho_\gamma^{(1)}(2)\, \frac{\partial f_{\alpha\gamma}(r_{12})}{\partial u_{\alpha;}(r_{12})}
$$

$$
\times \int C\{\mathbf{n}-\alpha,\gamma\}\rho_3^{(1)}(\{3\}) \cdots \rho_{n\nu}^{(1)}(\{n_\nu\})\, d\{\mathbf{n}-\alpha,\gamma\} \tag{40}
$$

with $C\{0\} = 1$. The quantity $C\{\mathbf{n}-\alpha,\gamma\}$ defines a sum of products of bonds associating $(\mathbf{n}-\alpha-\gamma)$ vertices with respectively at both ends α and the particle of species γ, both particles being separated by r_{12} and not directly connected.* The first summation in (40) is performed over the species $1,\ldots,\nu$. On the basis of previous definitions, the cluster function takes the form

$$
f_{\alpha\gamma}(r_{12}) = \left\{\exp\left(-\frac{u_{\alpha\gamma}(r_{12})}{kT}\right) - 1\right\}
$$

$$
= \left\{\exp\left[-\frac{1}{kT}\left(\frac{e_\alpha e_\gamma}{\epsilon_1}\gamma(r_{12}) + u_{\alpha\gamma}^*(r_{12})\right)\right] - 1\right\} \tag{41}
$$

in which the direct potential $u_{\alpha\gamma}$ is easily identifiable with (13), although the reciprocal of the distance is now substituted by a more general function $\gamma(r_{12})$, consistent with the previous remarks in Section 1 about electric reflection in the dielectric boundaries [cf. equations (6) and (7) in Section 1].

Assuming that $u_{\alpha\gamma}^*$ is independent of the charge e_α, i.e., that there is no short-range interaction dependence on e_α as in equations (24) and (26), one readily deduces from (39), (40), and (41) on account of

$$
\frac{\partial}{\partial e_\alpha} = \sum_\gamma \frac{\partial u_{\alpha\gamma}}{\partial e_\alpha} \frac{\partial}{\partial u_{\alpha\gamma}}
$$

---

* As a result of the properties of $S\{n\}$, these graphs are at least single-connected in the free vertices and without articulation point.

that

$$\left(\frac{\partial \tilde{\mu}_\alpha(1)}{\partial e_\alpha}\right)_{T,\rho^{(1)},\mathbf{P}^{(1)},p} = \left(\frac{\partial \mu_\alpha{}^0(1)}{\partial e_\alpha}\right)_{T,\rho^{(1)},\mathbf{P},p} + \psi^{\text{ext}}(1) + \sum_{\gamma=1} \int dr_{12} \frac{e_\gamma}{\epsilon_1} \gamma(r_{12})$$

$$\times \left. \right\} \rho_\gamma^{(1)}(2) \exp\left[-\frac{u_{\alpha\gamma}(r_{12})}{kT}\right] \sum_{(\mathbf{n}-\alpha,\gamma)\geqslant 0} \frac{1}{[\mathbf{n}-\alpha,\gamma]!}$$

$$\times \int C\{\mathbf{n}-\alpha,\gamma\}\rho_3^{(1)}(\{3\}) \cdots \rho_{n_\nu}^{(1)}(\{n_\nu\})\, d\{\mathbf{n}-\alpha\gamma\}\left.\right\}$$

(42)

By comparing the factor between the large braces appearing in the integrand of (42) with the expression of a pair correlation function in terms of the cluster integral method [43,44], it may be concluded that in the case of an inhomogeneous system, this quantity expresses the generic probability to find a given particle γ at a distance r_{12} from α. Thus,

$$\rho_\gamma^{(1)}(2)G_{\alpha\gamma}(r_{12}) = \rho_\gamma^{(1)}(2)(f_{\alpha\gamma}(r_{12}) + 1) \sum_{(\mathbf{n}-\alpha,\gamma)\geqslant 0} \frac{1}{[\mathbf{n}-\alpha\gamma]!}$$

$$\times \int C\{\mathbf{n}-\alpha,\gamma\}\rho_3^{(1)}(\{3\}) \cdots \rho_{n_\nu}^{(1)}(\{n_\nu\})\, d\{\mathbf{n}-\alpha,\gamma\} \quad (43)$$

One may immediately assert that if the particle α is removed from (1), this probability is essentially just $\rho_\gamma^{(1)}(2)$; hence, $G_{\alpha\gamma}(r_{12})$ is equal to unity. Therefore, one is led to rewrite (42) in a more simple form. Hence, on account of the electroneutrality condition $\sum_\gamma e_\gamma \rho_\gamma' = 0$, equation (42) will be

$$\left(\frac{\partial \tilde{\mu}_\alpha(1)}{\partial e_\alpha}\right)_{T,\rho^{(1)},\mathbf{P},p} = F_\alpha(1) + \psi_{(1)}^{\text{ext}}$$

$$+ \sum_\gamma \int dr_{12} \frac{e_\gamma}{\epsilon} \gamma(r_{12})[\rho_\gamma^{(1)}(2) - \rho_\gamma']$$

$$+ \sum_\gamma \int dr_{12} \frac{e_\gamma}{\epsilon} \gamma(r_{12})\rho_\gamma^{(1)}(2)[G_{\alpha\gamma}(r_{12}) - 1] \quad (44)$$

where

$$\left(\frac{\partial \mu_\alpha{}^0(1)}{\partial e_\alpha}\right)_{T,\rho^{(1)},\mathbf{P},p} \equiv F_\alpha(1)$$

contains the self-image effect of α and $G_{\alpha\gamma}(r_{12})$ is given by relation (43).

In accordance with previous remarks (Section 1), one recognizes precisely in the sum of the second and third right-hand side term of (44) the average coulomb potential $\bar{\psi}(1)$ at (1). This potential is thus expressed in terms of $\psi^{\text{ext}}(1)$ and of the average coulomb potential due to the remaining charges under the condition that α has been removed from its place. Accordingly, $\bar{\psi}(1)$ is the solution of the Poisson equation

$$\nabla^2 \bar{\psi}(1) = -\frac{4\pi}{\epsilon} \sum_{\gamma=1}^{\nu} e_\gamma \rho_\gamma^{(1)}(1) \tag{45}$$

for which the boundary condition $\bar{\psi}(1) = 0$ is taken in the bulk phase.

Since the remaining fourth right-hand side term of (44) contributes to the value of the local potential whenever the presence of α at (1) affects the neighboring distribution of charges, it is tantamount to conclude that this term corresponds to short-range and self-atmosphere effects on particle α, i.e., $\bar{\psi}_\alpha^{\text{sa}}(1)$. This leads finally to replacing (44) by the expression

$$\left(\frac{\partial \tilde{\mu}_\alpha(1)}{\partial e_\alpha}\right)_{T, \rho^{(1)}, \mathbf{P}, p} = F_\alpha(1) + \bar{\psi}(1) + \bar{\psi}_\alpha^{\text{sa}}(1) \tag{46}$$

Let us reconsider the equation of Gibbs (38) and focus our interest on the property of integrability for $\tilde{g}$. In view of the fact that (38) has been derived at constant composition set $\underline{\rho}^{(1)}$, the curl condition leads to

$$(\rho_\alpha^{(1)}[v(\mathbf{r})] - \rho_\alpha') \left[\frac{\partial}{\partial e_\beta} \left(\frac{\partial \tilde{\mu}_\alpha[v(\mathbf{r})]}{\partial e_\alpha}\right)_{\mathbf{P}, T, \rho^{(1)} p}\right]_{\rho^{(1)}} =$$

$$(\rho_\beta^{(1)}[v(\mathbf{r})] - \rho_\beta') \left[\frac{\partial}{\partial e_\alpha} \left(\frac{\partial \tilde{\mu}_\beta[V(\mathbf{r})]}{\partial e_\beta}\right)_{\mathbf{P}, T, \rho^{(1)} p}\right]_{\rho^{(1)}} \tag{47}$$

We might remark that a computation of local thermodynamic functions from statistical mechanical quantities would require an average of the previous quantities over the volume element $v(\mathbf{r})$. However, the electrochemical potential defined at some point of coordinate included in $v(r)$ will be equal to its average value $\tilde{\mu}_\alpha[v(\mathbf{r})]$ on account of the local and microscopic equilibrium condition. Furthermore, it is obvious that

$$\rho_\alpha^{(1)}[v(\mathbf{r})] = \int_{v(\mathbf{r})} \rho_\alpha^{(1)}(\{\alpha\}) \, d\{\alpha\} \tag{48}$$

so that finally (38) may be expressed in an equivalent unaveraged form. Hence, for the above considered particles α and β one gets from (46) the condition

$$(\rho_\alpha^{(1)}(1) - \rho_\alpha') \left[\frac{\partial}{\partial e_\beta} (\bar{\psi}(1) + \bar{\psi}_\alpha^{sa}(1)) \right]_{\rho^{(1)}}$$

$$= (\rho_\beta^{(1)}(1) - \rho_\beta') \left[\frac{\partial}{\partial e_\alpha} (\bar{\psi}(1) + \bar{\psi}_\beta^{sa}(1)) \right]_{\rho^{(1)}} \tag{49}$$

Here F_α has been assumed to be independent of the charge e_β .

3.4.2. *The Solution of Gouy and Chapman.* The viewpoint adopted separately by Gouy [21] and Chapman [22] involves the determination of the electrostatic average potential in the diffuse layer from the equation of Poisson, assuming a one-dimensional (planar electrode) and continuous system with constant permittivity ϵ_1 ,

$$\frac{\partial^2 \bar{\psi}(z)}{\partial z^2} = - \frac{4\pi}{\epsilon_1} \sum_{\gamma=1} e_\gamma \rho_\gamma^{(1)}(z) \tag{50}$$

It is postulated furthermore that

$$\bar{\psi}(z) > F_\alpha(z) + \bar{\psi}_\alpha^{sa}(z) \tag{51}$$

so that the expression for the singlet distribution function is given by

$$\rho_\gamma^{(1)}(z) = \rho_\gamma' \exp\left(- \frac{e_\gamma \bar{\psi}(z)}{kT} \right) \tag{52}$$

A first integration of the Poisson–Boltzmann equations (50) and (52) may now be performed under the following conditions:

$$[\bar{\psi}(z)]_{z \to \infty} = 0 \qquad \left[\frac{d\bar{\psi}(z)}{dz} \right]_{z \to \infty} = 0 \tag{53}$$

and in the case of a symmetric $e_i - e_i$ electrolyte leads to the value

$$\left[\frac{d\bar{\psi}(z)}{dz} \right]_z = \frac{2kT}{e_i} \kappa_3 \sinh \frac{e_i \bar{\psi}(z)}{2kT} \tag{54}$$

A second integration of (50) and (52) yields

$$|\bar{\psi}(z)| = \frac{2kT}{|e_i|} \ln \coth \frac{1}{2} [(z - z_d)\kappa_3 + a_d] \tag{55}$$

where

$$a_d = \ln \coth \frac{|e_i|\, \bar{\psi}(z_d)}{4kT} \tag{56}$$

If one examines briefly the salient features of the Gouy–Chapman treatment, one may assert in the first place that this treatment corresponds to assuming that $\bar{\psi}(z)$, the Gouy potential, constitutes a solution of the diffuse-layer problem. Then, as in the case of Debye–Hückel, this statement is confirmed by the use of the Poisson–Boltzmann equation. Therefore, there is no need at any time to explicate the Gouy potential in terms of a superposition of coulombic bonds, as shown in (44). Nevertheless, it must be realized that through the application of Poisson's equation these bonds are implicitly present and thus the requirement of linear superposition of fields must always be satisfied. On the other hand, the interaction with induced surface charges is implied in the definition of the boundary value $\bar{\psi}(z_d)$ and in

$$\left(\frac{d\bar{\psi}(z)}{dz}\right)_{z_d} = -\frac{4\pi(q_a + q^m)}{\epsilon_1} \tag{57}$$

Yet relation (51) allows us to state that self-image terms have been neglected.

Consequently, one criterion of reasonableness of the Gouy–Chapman solution may be obtained by invoking the necessary properties of superposition of coulomb potentials [cf. equations (44) and (46)] and of integrability of (49). In the light of these two conditions, and within the limit of the Gouy–Chapman model, the following expression must be satisfied;

$$\int d\{\alpha\}\gamma(r_{1\alpha})(\rho_\alpha^{(1)}(\{\alpha\}) - \rho_\alpha')(\rho_\beta^{(1)}(1) - \rho_\beta')$$

$$= \int d\{\beta\}\gamma(r_{1\beta})(\rho_\beta^{(1)}(\{\beta\}) - \rho_\beta')(\rho_\alpha^{(1)}(1) - \rho_\alpha') \tag{58}$$

After introduction of the Gouy–Chapman distribution function (52), the condition (58) is expected to be satisfied only if a linear expansion of (52) in terms of the Gouy potential is performed. In this case we are led to

$$(\rho_\beta^{(1)}(2) - \rho_\beta')(\rho_\alpha^{(1)}(1) - \rho_\alpha') = (\rho_\alpha^{(1)}(2) - \rho_\alpha')(\rho_\beta^{(1)}(1) - \rho_\beta')$$

$$= \rho_\alpha'\rho_\beta' \frac{e_\alpha e_\beta}{(kT)^2}\, \bar{\psi}(1)\bar{\psi}(2) \tag{59}$$

By considering the linearization condition and the approximation (51) it is easy to deduce that the Gouy potential must obey the relations

$$I_5 = \left| \frac{e_\alpha \bar{\psi}(z)}{kT} \right| \ll 1 \tag{60}$$

$$\left| \frac{e_\alpha}{2\epsilon_1 z' \bar{\psi}(z)} \right| \ll 1 \tag{61}$$

where z' is the virtual distance defined in Section 1, equation (9). Finally, it remains to introduce the condition

$$\bar{\psi}_\alpha^{\text{sa}}(z) \ll \bar{\psi}(z) \tag{62}$$

which will be expressed more explicitly in the next section.

According to the restrictions given above, it is surprising that the solution of the nonlinearized Gouy–Chapman treatment appears to give some significant interpretations of electrokinetic processes and double-layer capacity measurements. This cannot be theoretically explained in the frame of the above theory and, in a sense, may seem to be fortuitous.

3.4.3. *Determination of the Self-Atmosphere Effect by Means of the Cluster Expansion Method.* Mayer [28], in a now-familiar method, has derived the Debye–Hückel limiting law by using the cluster method. An outline of the basic aspects of this derivation will suffice here. The first step is to linearize the cluster function (41) with respect to the electrostatic part $(e_\alpha e_\beta / \epsilon_1) \gamma$ and to drop the short-range contribution. According to this procedure only products of $(e_\alpha e_\beta / \epsilon_1) \gamma$ bonds will appear in the cluster integral. The next operation is to retain in the sum of cluster integrals only those terms corresponding to single cyclic graphs.

Stillinger and Buff [4] solved equations (39), (43), and (44) in the case of a very dilute diffuse layer on the same basis. They also assumed a linearized ring cluster approximation for the system of ions adsorbed in the diffuse layer so that for a given set of ions, only the mutual interaction of the lowest order in γ is represented. However, as was established in Section 1, the dielectric discontinuities at $z = 0$ and $z = h_d$ affect the potential, which is now given by the expression of Section 1, equation (8). For the same reason, the mutual effect between pairs of ions, one of which is adsorbed at the inner Helmholtz plane and the other in the diffuse layer, is described by the expression of

Section 1, equation (10). Since, in the dilute case, these two ions are at some distance apart, it has been further assumed that the interaction of ions belonging to phase I and II (cf. Fig. 1) can be restricted to a simple pairwise bond and that any contribution of graphs of higher order is negligible. On account of the previous assumptions, the electrochemical potential (39) may be written

$$\tilde{\mu}_\alpha(1) = \mu_\alpha{}^0(1) + e_\alpha \psi_{(1)}^{\text{ext}}$$

$$+ kT \sum_{\gamma=1}^{\nu} \int_{\text{I+II}} d\mathbf{r}_\gamma \frac{e_\alpha e_\gamma}{\epsilon_1} \gamma(r_\gamma)\rho_\gamma^{(1)}(\{\gamma\})$$

$$+ \sum_{[\mathbf{n}-\alpha]\geq 1} \frac{[\mathbf{n}-\alpha]!}{2[\mathbf{n}-\alpha]!} \int_{\text{II}} e_\alpha \psi_{\beta,\text{II}}^D(r_{1\beta}) \, e_\beta \psi_{\gamma,\text{II}}^D(r_{\gamma\beta}) \cdots e_{n_\nu}\psi_{n_\nu,\text{II}}^D(r_{n_\nu})$$

$$\times \rho_\beta^{(1)}(\{\beta\})\rho_\gamma^{(1)}(\{\gamma\}) \cdots \rho_{n_\nu}^{(1)}(\{n_\nu\}) \, d\{\mathbf{n}-\alpha\} + kT \ln \rho_\alpha^{(1)} \tag{63}$$

As previously stated, the second and third right-hand side terms of (63) yield the average electrostatic potential at (1), and the fourth term represents the self-atmosphere effect limited under the present conditions to interactions within the diffuse layer. Each integral of (63) corresponds to a ring graph of n particles. Taking into account the indiscernability one divides by $[\mathbf{n}-\alpha]!$. The factor $[\mathbf{n}-\alpha]!/2$ indicates the number of distinguishable permutations of numbered vertices belonging to a cycle. The integration of Buff and Stillinger leads to the expression

$$\tilde{\mu}_\alpha(z_1) = \mu_\alpha{}^0(z_1) + e_\alpha \bar{\psi}(z_1)$$

$$- \frac{e_\alpha{}^2}{4\epsilon_1 z_1} \exp[-2\overline{\kappa_3(z_1)}\, z_1] - \frac{e_\alpha{}^2}{2\epsilon_1} \overline{\kappa_3(z_1)} + kT \ln \rho_\alpha^{(1)} \tag{64}$$

where z_1 is the linear coordinate, normal to the electrode, at the position of α and $\overline{\kappa_3(z_1)}$ represents the average local value of the Debye–Hückel parameter (37); hence,

$$\overline{\kappa_3(z_1)} = \lim_{z_1 \to z_2} \frac{1}{z_1 - z_2} \int_{z_1}^{z_2} \kappa_3(z) \, dz \tag{65}$$

The two last terms of (64) are, respectively, the ions own shielded image term and the local average activity correction at position z_1. The former effect is comparatively small, in most cases, whenever the condition $2z_1\overline{\kappa_3(z_1)} > 1$ is fulfilled. According to the equilibrium condition of Section 2, equation (26), and the effect of polarization in the external field, the singlet distribution function $\rho_\alpha^{(1)}(z_1)$ can thus be written as given in Section 2, equation (27), where

$$kT \ln f_{s0}(z_1) = \frac{e_\alpha \bar{\psi}_\alpha^{as}}{2} (z_1) \tag{66}$$

hence

$$kT \ln f_{(s0)}(z_1) = -\frac{e_\alpha^2}{2\epsilon_1} \left\{ \overline{\kappa_3(z_1)} + \frac{1}{2z_1} \exp[-2\overline{\kappa_3(z_1)}\, z_1] \right\} \tag{67}$$

An important aspect of the comparison between Section 2, equation (12), and (64) or Section 2, equation (27), and (66) and (67) has been stated by Buff and Stillinger [4]. Thus, it may be observed that "in spite of the fact that the activity correction for an ion at z_1 is due to its average charge cloud extending over distances comparable with the entire diffuse layer thickness, in dilute solution limit the pointwise use of bulk solution activity coefficient, as suggested by the local thermodynamic balance, is rigorously valid."

However, in the general demonstration of (64), which has been described here in its principles, no attempt was made by Buff and Stillinger to evaluate the incidence of a nonlinear and composition-dependent dielectric behavior. The way in which short-range effects could be inserted into relation (64) was only formally emphasized. It has been stressed that the potential $\bar{\psi}(z_1)$ in Section 2, equation (27), should be replaced by $\bar{\psi}_{cav}(z_1)$ differing from that used previously in that the associated charge distribution producing it is the mean double-layer charge minus the charge that would lie on the average at the site of adsorption of the finite size ion (the so-called cavity field effect). The particular incidence of finite ionic volumes will be reconsidered next, on a more consistent basis, by means of the treatment of Stillinger and Kirkwood [45].

3.4.4. *The Diffuse Layer Version of the Expansion Method in Terms of a Charging Parameter.* Let us assume that the system consists of **n** particles of species 1, 2,..., some of which are electrostatically charged. These particles are subjected to forces arising from the presence of a charged planar interface as well as from the presence of other particles. Hence, the total direct potential of interaction $U(\mathbf{n})$ may be split into singlet contributions resulting from the interface and pair contributions attributable to the particle interactions

$$U(\{\mathbf{n}\}) = \sum_{\{n\}} U^{(1)}(\{i\}) + \sum_{\substack{i \neq j \\ \{n\}}} U^{(2)}(\{i,j\}) \tag{68}$$

The singlet and pair potentials may further be separated into parts of pure electrostatic and pure short-range character. The pure electrostatic interaction arises as in the method of Kirkwood and Poirier [27] for an assembly of ions, each of which may carry a fraction ξ_1, ξ_2,..., ξ_μ,..., ξ_ν of its charge (for neutral solvent species these parameters vanish). Thus, we may write the total direct potential for a system in which a selected particle α situated at $\{\alpha\}$ carries ξ_α of its charge e_α:

$$U(\{\mathbf{n}\}) = U^0(\{\mathbf{n}\}) + \xi_\alpha \varphi_\alpha(\{\alpha\})$$

$$= U^0(\{\mathbf{n}\}) + \xi_\alpha \left\{ V^{(1)c}(\{\alpha\}) + \sum_{\{n\}, j \neq \alpha} \xi_j V^{(2)c}(\{\alpha, j\}) \right\} \qquad (69)$$

where $U^0(\mathbf{n})$ represents the interaction potential for $\xi_\alpha = 0$ and $V^{(1)c}$ and $V^{(2)c}$ are the pure electrostatic part of respectively the singlet and pairwise interaction potential, $U^{(1)}(\{i\})$ and $U^{(2)}(\{ij\})$, corresponding to α. Hence, at the plane electrode interphase in the absence of specific adsorption,

$$V^{(1)c}(\{\alpha\}) = -2\pi q^m e_\alpha \left\{ \frac{h_d}{\epsilon_0} + \frac{z_\alpha - h_d}{\epsilon_1} \right\} + A \qquad (70)$$

$$V^{(2)c}(\{\alpha, j\}) = \frac{e_\alpha e_j}{\epsilon_1 r_{\alpha j}} \qquad (71)$$

The constant A depends on the choice of the zero of the electrostatic energy and will not affect the result due to the fact that it cancels out during the treatment. In a similar way to (69), the singlet potential of average force $W_\alpha^{(1)}$ is given by

$$W_\alpha^{(1)}(\{\alpha\}) = W_\alpha^{(1)0}(\{\alpha\}) + \xi_\alpha e_\alpha \Psi(\{\alpha\}) \qquad (72)$$

The first of these terms $W_\alpha^{(1)0}(\{\alpha\})$ corresponds to a system where $\xi_\alpha = 0$, and the second term contains the total number of electrostatic interactions in the diffuse layer appearing when particle α is charged to the amount ξ_α. Thus, the function $\Psi(\{\alpha\})$ introduced in (72) acts as an electrostatic average potential at $\{\alpha\}$. On ground of the definition of the singlet potential of average force (12) and (16) and the superposition approximation, one may infer that the second right-hand side term of (72) results from the cumulative action of (1) the potential $V^{(1)c}(\{\alpha\})$ due to the external field and (2) the statistical average over the entire system with $\{\alpha\}$ excluded of $V^{(2)c}(\{\alpha, j\})$. Thus, to within a constant independent of $\{\alpha\}$, equation (72) may be rewritten under the form

$$W_\alpha^{(1)}(\{\alpha\}) - W_\alpha^{(1)0}(\{\alpha\}) = \xi_\alpha \frac{\int \exp[-U(\{\mathbf{n}\})/kT] \cdot \varphi_\alpha(\{\alpha\}) \, d\{\mathbf{n} - \alpha\}}{\int \exp[-U(\{\mathbf{n}\})/kT] \, d\{\mathbf{n} - \alpha\}} + \text{const} \tag{73}$$

Now, if the value (69) is introduced into expression (73) and if the exponentials under the integrals are expanded in function of ξ_α, it may readily be shown that (73) transforms into

$$W_\alpha^{(1)}(\{\alpha\}) - W_\alpha^{(1)0}(\{\alpha\})$$

$$= \xi_\alpha \frac{\int \exp[-U^0(\{\mathbf{n}\})/kT] \, \varphi_\alpha(\{\alpha\}) \, d\{\mathbf{n} - \alpha\}}{\int \exp[-U^0(\{\mathbf{n}\})/kT] \, d\{\mathbf{n} - \alpha\}} + \text{const}$$

$$= \xi_\alpha \left[V^{(1)c}(\{\alpha\}) + \sum_{\{\mathbf{n}-\alpha\}<j} \xi_j \frac{\int \exp[-U^0(\mathbf{n})/kT] \, V^{(2)c}(\{\alpha, j\}) \, d(\mathbf{n} - \alpha\}}{\int \exp[-U^0(\{\mathbf{n}\})/kT] \, d\{\mathbf{n} - \alpha\}} \right]$$

$$+ \text{const} \tag{74}$$

where the point of view has been taken to retain only the linear term into ξ_α, similar to what has been done in the treatment of Debye–Hückel's limiting law [27]. Stillinger and Kirkwood [45] have derived expression (74) by means of a number of rigorous mathematical manipulations associated with the leading principle of linearization into the parameter ξ_α. Considering the fact that the singlet potential $V^{(1)c}(\{\alpha\})$ varies linearly with the coordinate z_α, these authors have also indicated that the statistical average of $V^{(1)c}(\{\alpha\})$ depends only on z_d as a result of the planar symmetry of the double layer system. Therefore, taking into account the relations (70), (71), and (72) and definitions (12) and (16), a twofold differentiation of (74) with regard to z_α, yields

$$\frac{d^2 \Psi(z_\alpha)}{dz_\alpha^2} = \frac{d^2}{dz_\alpha^2} \frac{1}{e_\alpha}$$

$$\times \frac{\int \exp -\dfrac{U^0(\{\mathbf{n}\})}{kT} \sum_{\{\mathbf{n}-\alpha\}<j} \xi_j V^{(2)c}(\{\alpha, j\}) \, d\{\mathbf{n} - \alpha\} \int \exp -\dfrac{U^0(\{\mathbf{n}\})}{kT} d\{\mathbf{n}\}}{\int \exp -\dfrac{U^0(\{\mathbf{n}\})}{kT} d\{\mathbf{n}\} \int \exp -\dfrac{U^0(\{\mathbf{n}\})}{kT} d\{\mathbf{n} - \alpha\}}$$

$$= \frac{1}{\epsilon_1} \frac{d^2}{dz_\alpha^2} \int \frac{dr_{\alpha 2}}{r_{\alpha 2}} \left[\rho_+ \xi_+ e_+ \frac{g_{\alpha+}^{(2)0}(\{\alpha, 2\})}{g^{(1)0}(\{\alpha\})} + \rho_- \xi_- e_- \frac{g_{\alpha-}^{(2)0}(\{\alpha, 2\})}{g_\alpha^{(1)0}(\{\alpha\})} \right] \tag{75}$$

where $g_\alpha^{(1)0}(\{\alpha\})$ and $g_{\alpha\beta}^{(2)0}(\{\alpha, \beta\})$ are respectively the singlet and pair correlation function for the discharged particle α and where the occupation of site of coordinate $\{2\}$ is considered for both ionic species present in the system. Since the electrode surface behaves as a rigid wall, the singlet correlation function $g_\alpha^{(1)0}(\{\alpha\})$ settles down rapidly to unity at a distance longer than z_d (cf. Fig. 1). In order to evaluate the pairwise correlation function $g_{\alpha\beta}^{(2)0}(\{\alpha, \beta\})$, one is led to introduce an approximation analogous to the superposition approximation used in the molecular theory of bulk liquids [cf. equation (19)]. Thus, it is assumed that the pairwise potential of mean force for the discharged particle α, $W_{\alpha\beta}^{(2)0}(\{\alpha, 2\})$, may be obtained in superposing the singlet potential of force of the ion located at $\{\beta\}$ with the bulk pair potential $W_{\alpha\beta}^{(2)0b}(\{\alpha, \beta\})$, which is realized whenever the external field is virtually switched off at $\{\beta\}$. Hence,

$$g_{\alpha\beta}^{(2)0}(\{\alpha, \beta\}) = \exp\left[-\frac{W_{\alpha\beta}^{(2)0}(\{\alpha, \beta\})}{kT}\right]$$

$$= \exp\left[-\frac{W_{\alpha\beta}^{(2)0b}(\{\alpha, \beta\}) + W_{\alpha\beta}^{(1)c}(\{\beta\})}{kT}\right]$$

$$= g_{\alpha\beta}^{(2)0b}(\{\alpha, \beta\})g_\beta^{(1)}(\{\beta\}) \tag{76}$$

If now, on account of (16) and (72), the singlet function $g_\beta^{(1)}(\{\beta\})$ is linearly expanded in terms of the charging parameter ξ_β, and if $g_\beta^{(1)0}(\{\beta\})$ is unity as in the case for $g_\alpha^{(1)0}(\{\alpha\})$, it appears that

$$g_\beta^{(1)}(\{\beta\}) = g_\beta^{(1)0}\left(1 - \xi_\beta \frac{e_\beta}{kT} \Psi(z_\beta)\right)$$

$$= 1 - \xi_\beta \frac{e_\beta}{kT} \Psi(z_\beta) \tag{77}$$

By inserting (76) and (77) into the integrand of (75), a first term will appear containing exclusively $g_{\alpha\beta}^{(2)0b}(\{\alpha, \beta\})$ and, therefore, being essentially independent of the coordinate $z_\alpha \geq z_d$. As a result of the value of the remaining term, equation (75) transforms into

$$\frac{d^2\Psi(z_\alpha)}{dz_\alpha{}^2} = -\frac{1}{\epsilon_1 kT} \frac{d^2}{dz_\alpha{}^2} \int \frac{d\mathbf{r}_{\alpha2}}{r_{\alpha2}} \left[\rho_+(\xi_+e_+)^2 g_{\alpha+}^{(2)0b}(\{\alpha, 2\})\Psi(z_2)\right.$$

$$\left. + \rho_-(\xi_-e_-)^2 g_{\alpha-}^{(2)0b}(\{\alpha, 2\})\Psi(z_2)\right] \tag{78}$$

Thus, for a single-component salt solution of symmetric, uniformly charged spherical particles suspended in a dielectric continuum, the differential equation (78) leads to

$$\frac{d^2\Psi(z_\alpha)}{dz_\alpha^2} = -\frac{\kappa_3^2}{4\pi}\frac{d^2}{dz_\alpha^2}\int \frac{d\mathbf{r}_{\alpha\beta}}{r_{\alpha\beta}}\, g_{\alpha\beta}^{(2)0b}(\{\alpha, \beta\})\Psi(z_\beta) \tag{79}$$

It is now necessary to introduce cylindrical coordinates $r_{\alpha\beta}^2 = s_{\alpha\beta}^2 + (z_\alpha - z_\beta)^2$ and to integrate equation (79) over all values of $s_{\alpha\beta}$. During this operation it should be remembered that $\partial g_{\alpha\beta}^{(2)0b}(\{\alpha, \beta\})/\partial r_{\alpha\beta}$ is just zero except for $r_{\alpha\beta}$ being very small. Hence, it follows that for $z_\alpha \geqslant z_d$ and equal-sized ions, (79) yields

$$\frac{d^2\Psi(z_\alpha)}{dz_\alpha^2} = -\frac{\kappa_3^2}{2}\int_{z_d}^{\infty}\left[\frac{d}{dz_\alpha}g_{\alpha\beta}^{(2)0b}(z_\alpha - z_\beta)\right]\Psi(z_\beta)\,dz_\beta \tag{80}$$

According to the model of hard spheres exhibiting an inpenetrability distance of a, one has

$$g_{\alpha\beta}^{(2)0b}(z_\alpha - z_\beta) = \begin{cases} 0 & 0 < |z_\alpha - z_\beta| < a \\ 1 & |z_\alpha - z_\beta| > a \end{cases} \tag{81}$$

The bulk correlation function may be considered as a function of the absolute value of the distance $|z_\alpha - z_\beta|$, so that for a given z_β, the kernel

$$\left[\frac{d}{dz_\alpha}g_{\alpha\beta}^{(2)0b}(z_\alpha - z_\beta)\right]$$

of (80) would be an even function of the distance. Since the derivative of this step function is a Dirac delta function, the integrodifferential equation (80) leads to

$$\frac{d^2\Psi(z_\alpha)}{dz_\alpha^2} = \begin{cases} 0 & z_\alpha < z_d - a & (82a) \\ \frac{1}{2}\kappa_3^2\Psi(z_\alpha + a) & z_d - a < z_\alpha < z_d + a & (82b) \\ \frac{1}{2}\kappa_3^2(\Psi(z_\alpha - a) + \Psi(z_\alpha + a)) & z_d + a < z_\alpha & (82c) \end{cases}$$

Thus, it appears that the model of hard spheres introduces in the right-hand side term of (82c) an average value of Ψ extending over the distance of a around z_α. Such a conclusion, of course, is valid for ionic densities far less than the corresponding close-packed situation.

For $a \to 0$, (82) leads obviously to an equation similar to the linearized Gouy–Chapman one, and a behavior of the form $\exp(-\kappa_3 z_\alpha)$ for $\Psi(z_\alpha)$ would be expected. In the case where a does not vanish, the solution of (82) becomes a sum of damped exponentials

$$\Psi(z_\alpha) = A_1 \exp\left[-Y_1\left(\frac{z_\alpha}{a}\right)\right] + A_2 \exp\left[-Y_2\frac{(z_\alpha)}{a}\right] \tag{83}$$

in which Y_1 and Y_2 are complex conjugate functions. This solution has been shown to give a damped sinusoid for $\kappa_3 a \geqslant 1.03$, thus implicating that $\Psi(z_\alpha)$ takes alternatively a positive and negative value along z. Naturally, the charge densities in the diffuse layer would likewise change from positive to negative values along z, so that the final structure of the diffuse layer would seem to correspond to alternative positive and negative sheets of charges ordered parallel to the electrode surface. This kind of arrangement, controlled by the effect of short-range interactions, is similar in some aspects, as pointed out by Stillinger and Kirkwood [45], to the lattice structure of ionic crystals.

3.4.5. *The Evaluation of the Electrostatic Potentials in the Diffuse Layer by Means of Stillinger and Kirkwood's Method.* In the preceding section we outlined the general treatment for the singlet potential of average force. As a result of this method, the singlet correlation function may be expressed in the following way:

$$g_\alpha^{(1)}(\{\alpha\}) = 1 - \xi_\alpha \frac{e_\alpha}{kT} \Psi(\{\alpha\}) \tag{84}$$

It has already been recognized that $\Psi(\{\alpha\})$ carries all coulombic effects acting on the ion α under consideration and depends, furthermore, on the short-range interaction model that was introduced in (81). From equations (12), (16), and (73), and in relation with (44) and (46), it is readily inferred that

$$\Psi(\{\alpha\}) = \bar{\psi}(\{\alpha\}) + \bar{\psi}_\alpha^{sa}(\{\alpha\}) \tag{85}$$

where $\bar{\psi}(\{\alpha\})$ is the electrostatic potential, and $\bar{\psi}_\alpha^{sa}$ represents just the contribution of short-range interactions since the linear approximation in (74) does not allow to deal with self-ionic atmosphere effects. According to (84), the Poisson equation takes the form

$$\nabla^2 \bar{\psi}(\{\alpha\}) = -\kappa_3^2 \Psi(\{\alpha\}) \tag{86}$$

Before proceeding further and solving equation (86), we wish to reexamine the basis of Stillinger and Kirkwood's treatment with a view toward assessing its applicability. Such analysis has been given by Krylov and Levich [5]. These authors have emphasized the fact that the correctness of (74) and consequently of (84) is determined within an accuracy of the first order in $\xi_\alpha \varphi_\alpha(\{\alpha\})/kT$. Accordingly, one must have the condition

$$\varphi_\alpha(\{\alpha\})/kT < 1$$

Unless this inequality is satisfied, the linearization of (74) in terms of $\varphi_\alpha(\{\alpha\})$ cannot reasonably be postulated. It is thus rather instructive to bring out the domain of validity of the method in connection with such an assumption. Therefore, let us follow the analysis given by Krylov and Levich [5].

It should be remarked first that the singlet potential of force has been derived under the supposition that the system is far from the close-packing state, so that the condition (32), $I_1 < 1$, has to be fulfilled. Moreover, the direct coulombic action exerted by α on its surrounding reaches a maximum, as far as neighboring ions are concerned, at the minimum possible value of $r_{\alpha\gamma}$, so that in the bulk of the solution the inequality (34), $I_3 < 1$, must be realized. From both conditions (32) and (34) it may be inferred that the relation

$$I_6 = \left| \frac{e_\alpha e_\gamma}{\epsilon_1 kT} \right| (\rho')^{1/3} \ll 1 \qquad (87)^*$$

holds in the bulk of the solution. According to the fact that $(\rho')^{-1/3}$ represents approximately the distance from α at which the probability of finding an ion γ of opposite charge is a maximum [26], it may further be concluded that the electrostatic energy of any ion in the bulk may be estimated by means of the relation

$$\frac{\varphi_\alpha(\{\alpha\})}{kT} \leqslant I_6 \qquad (88)$$

In the region of the diffuse layer, the external potential ψ^{ext} introduces itself with a maximum at the boundary z_α so that on account of (60) the value of $\varphi_\alpha(\{\alpha\})/kT$ is of the order of magnitude of I_5. It is now possible to estimate the conditions $I_5 \approx I_6 < 1$ for an aqueous univalent electrolyte solution at $300°K$ with $a = 4.2Å$; thus, $I_6 \simeq 0.6(c')^{1/3}$,

* $\rho' = \rho'_\alpha = \rho'_\gamma$ for a symmetric electrolyte.

c' being the molar bulk concentration per liter. Considering that c' must be smaller than 4.6 moles/liter, Krylov and Levich [5] went further in comparing this limit with the limit of the classical treatment of Debye–Hückel. In the Debye–Hückel treatment, the pair correlation function is linearly expanded in terms of I_2 [equation (33)], and it is known that condition (33) is fulfilled whenever $\kappa_3{}^3(\rho')^{-1} \simeq 54\sqrt{c'} < 1$ is realized. Such condition leads, for the case considered here, to $c' < 3.3\ 10^{-4}$ moles/liter. Hence, it finally may be concluded that the linear expansion of the singlet function (84), within an accuracy determined by the first order in I_6 and I_5, holds for values of c' at least two or three orders of magnitude larger than the values of c' that satisfy the usual linearization of the pair function in Debye–Hückel's treatment.

Let us return now to the system of equations (83) and (86). As mentioned earlier, Stillinger and Kirkwood considered a system without specific adsorption. In this case, the evaluation of the constants A_1 and A_2 in (83) requires the two following conditions: First, equation (82b) has to be satisfied at $z_\alpha = z_d$, hence,

$$\left[\frac{d^2\Psi(z_\alpha)}{dz_\alpha{}^2} = \frac{1}{2}\kappa_3{}^2(z_\alpha + a)\right]_{z_\alpha = z_d} \tag{89}$$

The second condition consists in the demand that the average charge of the diffuse layer must be completely neutralized by the charge density q^m; thus,

$$-q^m = +\int_{z_d}^{\infty} \sum_{\beta=1}^{\infty} \xi_\beta e_\beta \rho_\beta' g_\beta{}^{(1)}(z)\, dz \tag{90}$$

This condition, on account of (84), yields

$$-q^m = -\frac{\epsilon_1\kappa_3{}^2}{4\pi}\int_{z_d}^{\infty} \Psi(z)\, dz \tag{91}$$

The potential $\bar{\psi}(z_\alpha)$ at the outer Helmholtz plane is then computed by means of the classical electrostatic theory, using Section 1, equation (17), and (84). Accordingly,

$$\bar{\psi}(z_d) = \kappa_3{}^2 \int_{z_d}^{\infty} (z - z_d)\Psi(z)\, dz \tag{92}$$

The final solution for $\bar{\psi}(z_d)$ is of the form

$$\bar{\psi}(z_d) = \frac{4\pi q^m}{\kappa_3 \epsilon_1} (1 - \tfrac{1}{4}(\kappa_3 a)^2) \tag{93}$$

if $(\kappa_3 a) < 1$. For $\kappa_3 a \geqslant 1.03$, the solution becomes

$$\bar{\psi}(z_d) = \frac{4\pi q^m}{\kappa_3 \epsilon_1} F(0) \tag{94}$$

where the function $F(0)$ has been calculated by Stillinger and Kirkwood. The behavior of the ratio $\bar{\psi}(z_d)/(4\pi q^m/\kappa_3 \epsilon_1)$ is reproduced in Fig. 2 for a given range of values of $(\kappa_3 a)$. From this plot it may be concluded that the ratio decreases with increasing $\kappa_3 a$, changing sign near $\kappa_3 a = 1.46$.

The morphology of the potential distribution $\bar{\psi}(z)$ in the diffuse layer has been obtained by Krylov and Levich [5], on the basis of Stillinger and Kirkwood's treatment. These authors have furthermore extended their calculation to the case of zero charge in the metal, $q^m = 0$, and specific adsorption at z_a. The usual double-layer model represented in Fig. 1 has been considered. In setting up their equations,

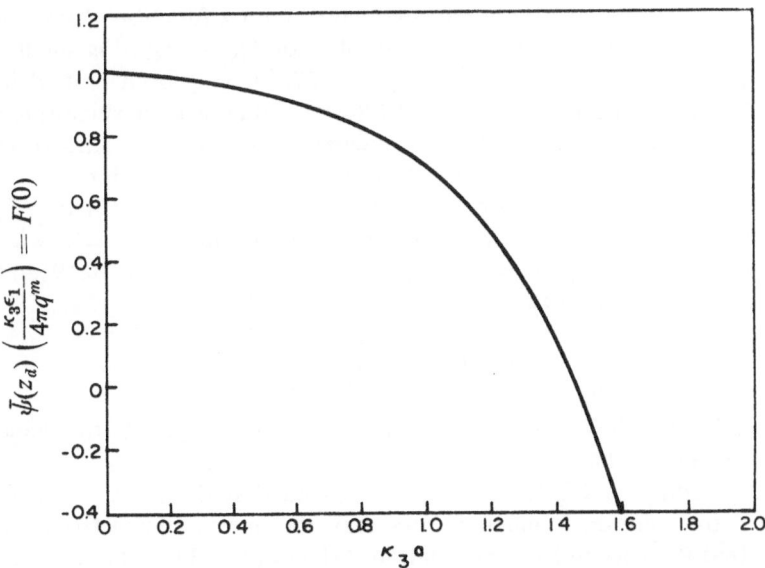

Fig. 2. The behavior of the Gouy potential $\bar{\psi}(z_d)$ as computed by Stillinger and Kirkwood [45] in function of $\kappa_3 a$.

Krylov and Levich have also postulated that the charge densities in the diffuse layer may be represented by the sum of densities induced separately by each adsorbed ion at z_a. With the preceding assumption, $\Psi(\{\alpha\})$ has been evaluated and inserted into (86). In order to solve the Poisson equation it is convenient to introduce Hänkel transforms. Such transforms lead to expressions for $\bar{\psi}(z)$ similar to those obtained in Section 1. The boundary conditions of Section 1, equations (4), also remain valid. The final potentials are averaged over the plane xy.

The average electrostatic potential in the diffuse layer corresponding to the condition $(\kappa_3 a) \ll 1$ is then described by

$$\bar{\psi}(z) = \frac{4\pi q_a}{\epsilon_1 \kappa_3} \exp\left[-\kappa_3 a\left(\frac{z - z_d}{a}\right)\right] \qquad (95)$$

which is simply the solution of the linearized Poisson–Boltzmann equation. In the case where $\kappa_3 a \geqslant 1.03$, the mathematical treatment leads to a much more involved solution of the form

$$\bar{\psi}(z) = \frac{4\pi q_a}{\epsilon_1 \kappa_3} \exp\left[-\alpha_1\left(\frac{z - z_d}{a}\right)\right] F\left(\frac{z - z_d}{a}\right) \qquad (96)$$

in which α_1 is the real part of the conjugate complex function Y already mentioned in (83) and where the multiplicator $F[(z - z_d)/a]$ is smaller than unity for $z_d \leqslant z \leqslant \infty$, the values of $F(0)$ having been plotted in Fig. 2. If in both relations (95) and (96) the value of q_a is substituted by q^m, one is led back to the case considered by Stillinger and Kirkwood of finite charge density q^m in the absence of specific adsorption.

The potential distribution in some systems has been evaluated [5] and is shown in Fig. 3. The values $\kappa_3 a = 1.03$ and $\kappa_3 a = 1.15$ with $a = 4.2$ Å and $T = 300°K$ correspond respectively to $c' = 0.59$ and 0.75 moles/liter. The behavior of

$$Z(x^*) = \frac{\bar{\psi}(z)}{4\pi q_{(a)}^{(m)}/\epsilon_1 \kappa_3}$$

in function of $x^* = (z - z_d)/a$ is compared in this figure to the value $\exp(-\kappa_3 a x^*)$ of the linearized Gouy–Chapman case.

According to Fig. 3, it may be inferred that the potential drop toward the solution is much steeper than the one predicted from the linearized Poisson–Boltzmann relation. When $\kappa_3 a = 1.03$, the function $Z(x^*)$ falls at $x^* = 3$ to 0.5% of its value at $x^* = 0$. This is smaller by a power of ten than the corresponding value of $\exp(-\kappa_3 a x^*)$.

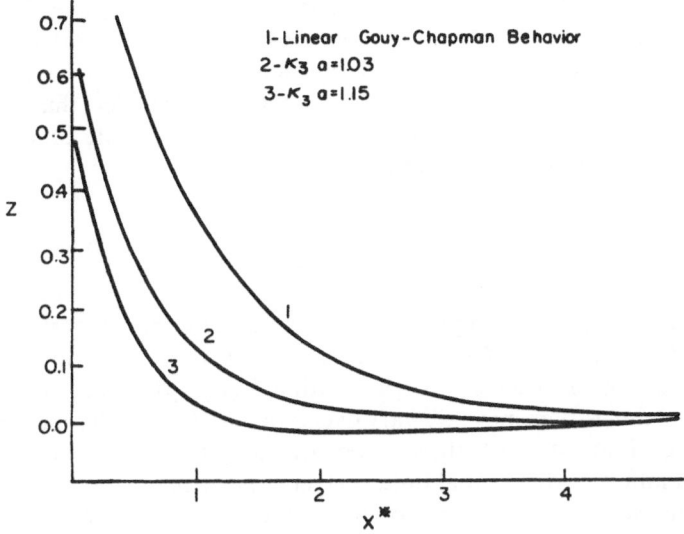

Fig. 3. The potential distribution in the diffuse layer as a function of $\kappa_3 a$
(cf. Krylov and Levich [5]).

When $\kappa_3 a = 1.15$, the function $Z(x^*)$ appears to oscillate weakly around zero for $x^* > 2$, but the deviation in the range $2 < x^* < 5$ is less than 0.7% of the value at $x^* = 0$.

Thus, one observes that the effective thickness of the diffuse layer is always smaller than $2/(\kappa_3)$ and that it decreases at a faster rate with increasing ionic strength. This appears precisely to justify the Ershler [6] approximation, which has been mentioned in relation with Section 1, equation (16).

Finally, it is worth emphasizing that the treatment of Stillinger and Kirkwood, in dealing with a strict superposition of effects, satisfies the integrability condition. This conclusion may readily be drawn from (49), (58), (59), (74), and (84).

Yet, the treament of Stillinger and Kirkwood does not yield the well-known hyperbolic behavior of the differential diffuse layer capacity, which has been experimentally established by Grahame [39] at low ionic strength. To clarify this problem we shall, in the following analysis, determine the limitations imposed on the values of the external field by the requirement that $I_5 < 1$ [cf.

equation (60)]. As noted earlier, the demand that $I_5 \simeq I_6 < 1$ limits the applicability of (93), (94), (95), and (96). According to (95) and (96), this condition can now be explicitly stated on the basis that the multiplicator of $4\pi q_{(a)}^{(m)}/\epsilon_1 \kappa_3$ in (95) and (96) is smaller than unity. Hence, as indicated by Krylov and Levich, one may write on account of $I_5 = I_6$ and (95) and (96) that

$$| q_{(a)}^{(m)} | \leqslant \frac{e_y \kappa_3 \rho'^{1/3}}{4\pi} \tag{97}$$

This relation leads for aqueous solutions of uni-univalent electrolyte to $| q_{(a)}^{(m)} | \leqslant 4.7(c')^{5/8}$ μCb/cm². Therefore, at $c' = 10^{-2}$ moles/liter, q_a or q^m is restricted to lie within ± 0.1 μCb/cm². Only for $c' = -2$ moles/liter may $q_{(a)}^{(m)}$ vary over a range of ± 1 μCb/cm². But at such high ionic strength there is no way to get any more a sensitive determination of differential diffuse layer capacities.

The strong limitations of the statistical approach plus the disconcerting fit between some experimental observations [39] and the nonlinearized Gouy–Chapman treatment have led different authors to consider some less rigorous procedures. One of these theories, based on the local thermodynamic formalism, is given next.

3.4.6. *Application of the Local Thermodynamic Method.* By application of the local thermodynamic method we succeeded in Section 2 in deriving a local distribution function [equation (27)] of solute particles in the diffuse layer. This formalism may lead to some predictions concerning the detailed structure of the diffuse layer if there is a clear way to express ϵ, the permittivity, as a function of the local concentration ρ_y and field E. Accordingly, we shall assume small changes of concentration ρ_y and ρ_1 of the solvent and relative small effects of the external field E. It is then possible to expand the permittivity linearly in terms of these quantities around ϵ_{10}, the permittivity of the pure solvent in the absence of an electric field [12]. Hence one has

$\epsilon(T, \rho_1 \ldots \rho_v, E^2)$

$$= \epsilon_{10} + \sum_v \left(\frac{\partial \epsilon}{\partial \rho_v} \right)_{\substack{T \\ \rho_1 \to \rho'_{10} \\ \rho_v \to 0 \\ E \to 0}} \rho_v + \left(\frac{\partial \epsilon}{\partial \rho_1} \right)_{\substack{T \\ \rho_1 \to \rho'_{10} \\ \rho_v \to 0 \\ E \to 0}} (\rho_1 - \rho'_{10}) + \left(\frac{\partial \epsilon}{\partial (E)^2} \right)_{\substack{T \\ \rho_1 \to \rho'_{10} \\ \rho_v \to 0 \\ E = 0}} E^2 \tag{98}$$

where ρ'_{10} indicates the bulk concentration of pure solvent. The square of E in (98) is justified by the argument that ϵ does not depend on the direction of the electrical field. The effect of dielectric saturation as reflected by the last term of (98) has been frequently discussed [35,46–48,89] and introduced in different approximate theories of the diffuse layer [38,34,89]. On the basis of some controversial measurements carried out by Malsch [49] in the year 1929 a value of $(\partial\epsilon/\partial(E)^2) \equiv -h = -3 \times 10^{-7}$ cm² $(V_{es})^{-2}$ has been established for fields less than $1.2\ 10^6$ V/cm. Accordingly, in the case of uni-univalent electrolyte, the expansion (98) yields

$$\epsilon(T, \rho_1 \ldots \rho_\nu, E^2) = \epsilon_{10} + b_+\rho_+ + b_-\rho_- + b_1(\rho_1 - \rho'_{10}) - hE^2 \quad (99)$$

where

$$b_- \equiv \left(\frac{\partial\epsilon}{\partial\rho_-}\right)_{T,\rho_+,\rho_1,E=0}$$

$$b_+ \equiv \left(\frac{\partial\epsilon}{\partial\rho_+}\right)_{T,\rho_-,\rho_1,E=0} \quad (100)$$

$$b_1 \equiv \left(\frac{\partial\epsilon}{\partial\rho_1}\right)_{T,\rho_-,\rho_+,E=0}$$

For strong, uni-univalent electrolyte, Hückel [50] and Sack [51] have suggested expressing the decrement of permittivity in the following way:

$$\epsilon(T, \rho_1, \ldots, \rho_\nu, 0) = \epsilon_{10} + \delta_+\rho_+ + \delta_-\rho_- \quad (101)$$

Later, Hasted et al. [46] carefully studied this relation and established the values of δ_+ and δ_-. They showed that over a range of concentration extending from 0.01 to 2 moles/liter these quantities are more or less constant and that their values are comprised between -3 and -20 liters/mole for a given number of ions.

The lowering of the static dielectric constant ϵ at $E = 0$ can be attributed to the following effects:

1. The "monopolization" of a given number of water dipoles in the ionic hydration sheaths. This effect would account for the fact that $|\delta_+| > |\delta_-|$, since the hindrance of rotation of water in the solvation sphere of anions is less than in the case of cations [52]. The relatively large values of δ_{OH-} and δ_{H+} may be related to the existence of a firmly associated group of water molecules around the considered H_3O^+ or OH^- ion [53].

2. A bond breaking caused by the ions and affecting the structure of bulk water [54].

3. The electrostriction of bulk water.

The two last effects are generally small (less than 10%) compared to the first.

It is now necessary to define the apparent molar volume $v_{\nu_0}^a$ of the solute species by means of the relation [55];

$$\rho_1 = \left(1 - \sum_\nu v_{\nu_0}^a \rho_\nu\right) \rho'_{10} \tag{102}$$

As a consequence of (99), (101), and (102), we have

$$b_+ = \delta_+ + v_{+0}^a \rho'_{10} b_1$$
$$b_- = \delta_- + v_{-0}^a \rho'_{10} b_1 \tag{103}$$

It should be remarked that the second right-hand side terms of both relations (103) account for the variation of water density that arises whenever a given particle of volume $v_{\gamma_0}^a$ is added at constant ρ_1 and $\rho_{\nu \neq \gamma}$. From recent measurements of the dielectric constant of water [14], it may be inferred that $\rho'_{10} b_1 \simeq 80$.

We thus dispose of the quantities b_1, h, $v_{+0}^a = v_{+0}^*$, $v_{-0}^a \simeq v_{-0}^*$ and may easily compute b_+, b_-, and ϵ. As a consequence, all quantities appearing in the exponent of Section 2, equation (27), are known. We may now couple these distribution functions to the one-dimensional Poisson equation of Section 2, equation (15); hence,

$$\frac{1}{\epsilon_{10}} \frac{d}{d\bar{\psi}(z)} \left(\epsilon \frac{d\bar{\psi}(z)}{dz}\right)^2 = \frac{kT\kappa_3^2 \epsilon}{e_\gamma} \left(\frac{\rho_-^{(1)}}{\rho'_-} - \frac{\rho_+^{(1)}}{\rho'_+}\right) \tag{104}$$

This relation, in conjunction with equation (57), yields the charge of the metal in the absence of specific adsorption. One readily finds

$$(q^m)^2 = \frac{\epsilon_{10} kT\kappa_3^2}{(4\pi)^2 e} \int_0^{\bar{\psi}(z_d)} \epsilon \left(\frac{\rho_-^{(1)}}{\rho'_-} - \frac{\rho_+^{(1)}}{\rho'_+}\right) d\bar{\psi}(z) \tag{105}$$

This expression has been integrated by iteration, assuming $f_{+0}/f'_{+0} = f_{-0}/f'_{-0} = $ constant near unity [12]. The solution is illustrated in Fig. 4, where the ratio of q^m to the charge q_{GC}^m anticipated from the Gouy–Chapman relation (54) is plotted as a function of $\bar{\psi}(z_d)$. The systems a, b, c, d, e, and f correspond to the values of the parameters given in Table I.

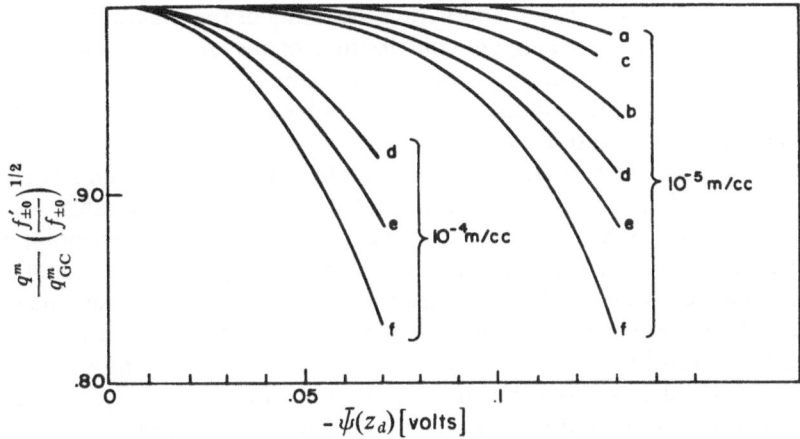

Fig. 4. The ratio of charge densities $\dfrac{q^m}{q^m_{GC}}\left(\dfrac{f'_{\pm 0}}{f_{\pm 0}}\right)^{1/2}$ as a function of the Gouy potential $\bar\psi(z_d)$ for the systems considered in Table I (cf. Hurwitz et al. [12]).

Table I.

	δ_+, cm³/mole	v^x_{+0}, cm³/mole	δ_-, cm³/mole	v^x_{-0}, cm³/mole
a	0	0	-5×10^3	-2.2
b	-3×10^3	40	-5×10^3	-2.2
c	0	40	-5×10^3	-2.2
d	-8×10^3	-1.55	-5×10^3	-2.2
e	-11×10^3	-1.09	-5×10^3	-2.2
f	-17×10^3	0.10	-5×10^3	-2.2

These systems have been chosen in order to illustrate on the cathodic side ($q^m < 0$), respectively, the influence:

1. Of the dielectric saturation ($b_+ = v^*_{+0} = 0$, $h \neq 0$): system a.
2. Of the dielectric saturation plus the electrostrictive pressure associated with a relatively large specific volume ($b_+ = 0$): system b.
3. Of the dielectric saturation, electrostrictive pressure, volume, increment of the dielectric constant ($b_+ > 0$): system e.
4. Of the dielectric saturation and the presumed specific influence: Na⁺, system d; Li⁺, system e; and H⁺, system f.

On account of Fig. 4 and within the limits of the present treatment one observes (1) a negligible contribution of the dielectric saturation, as has been already emphasized by Grahame [35], (2) a small effect of the specific volume, and (3) appreciable influences corresponding to the specific electrostriction parameters b_+.

Thus far the ratio of the activity coefficient $f'_{\gamma_0}/f_{\gamma_0}$ has been assumed to be equal to unity. In order to introduce a more reasonable value for this ratio, it is important to remember that these coefficients have been defined at zero electric field when local conditions of temperature and concentration are kept constant and identical to those existing under the external field. Furthermore we have indicated in (67) that at large dilution the use of bulk activities is valid.* In Fig. 4, we are, of course, already far from the condition of dilution agreeing with the derivation of equation (67). If, nevertheless, one assumes that the conclusion drawn from (67) still holds, it may be shown [12] that in the range of charge densities and potentials considered in Fig. 4, the effect of $f_{\gamma_0}/f'_{\gamma_0}$ is not appreciable, at least for systems d, e, and f.

The incidence of *ad hoc* corrections for f_{γ_0} at large concentration of finite size ions has been considered [28-34,38,56,57]. In a first approach to size effects, the cavity field can be isolated from f_{γ_0} and $\bar{\psi}_{cav}(z)$ retained in place of $\bar{\psi}(z)$ as indicated on page 179.† Unfortunately there is theoretical justification from Section 3.4.5 that size effects will affect the qualitative nature of f_{γ_0} more deeply.

3.5. General Evolution of Ideas on Specific Adsorption in the Inner Layer

It was recognized early that adsorption at the electrode cannot be interpreted solely in terms of the model of diffuse layer. The possibility of some more specific interaction with the surface was realized by Gouy [21] and Stern [58], who in 1924 treated a two-dimensional model of adsorption based on the Langmuir isotherm. Since then the strong dependence of most electrocapillary phenomena and electrode kinetics on specific adsorption has been proved experimentally and is theoretically accepted. Further explorations in this field have emerged from the existence of conceptual similarities between adsorption at the metal–gas and metal–solution interface, and from a considerably simpler situation on statistical mechanical ground, than for the diffuse

---

* Cf. also Williams [56] and Hill [57].
† Note added in proof: Bell and Levine [90] have considered such an effect recently.

layer. An adequate synthesis between the model of inner layer and diffuse layer has been achieved by Grahame [39] and resulted in successful verification of the existing double-layer theories. We refer to the review of Parsons [59] and Delahay [60] on this subject and intend to outline here only some of the important steps in the progress of basic concepts on this matter.

The model of Stern, although assuming independent localized sites of adsorption (Langmuir's type), from the electrostatic viewpoint, relies on a homogeneous double layer with charges smeared out over the Inner Helmholtz plane. Esin and Markov [61] pointed out later, on experimental evidence, some important inconsistency of this model. They advocated a possible influence of discreteness-of-charge effects. De Boer [62] and Langmuir [63] for the metal–gas interphase and Frumkin [64] at the electrode interface, had apparently been first to emphasize such effects.

The foregoing discreteness of charge consideration happened to be the essential background of an important number of works. Esin and Shikov [65] and Erschler [6] stressed the viewpoint that the charges are distributed at the lattice points of a hexagonal array. The model of Erschler included, furthermore, the approximation of an equipotential conducting plane instead of the diffuse layer. More recently, this model has been treated in a rigorous way [67,68] and has provided some interesting insight into the problem of double layer.

The main defect of the Ershler-type model is its inability to allow for thermal motion of adions (specifically adsorbed ions), thus reducing the ionic distribution function to its most extreme form, especially inadequate in the case of a liquid–metal substrate like mercury. In order to make allowance for such thermal motion, Levine et al. [69] suggested an excellent approach based on an idea given earlier by Grahame [70]. But it was not until the work of Buff and Stillinger [4] that the question was revived on sound statistical mechanical ground. The method suggested by these authors has provided the theoretical background for the treatment given below.

Observations of a large number of systems disclose [71,72] that the initial heat of adsorption, measured at zero charge of the electrode and vanishing coverage, may be related to significant change in hydration and polarizability of the particle attracted in the inner layer. In the present account, we confine our attention to the statistical mechanical aspects without going into a detailed analysis of the nature of adsorptive forces involved in the initial heat of adsorption. We further assume a

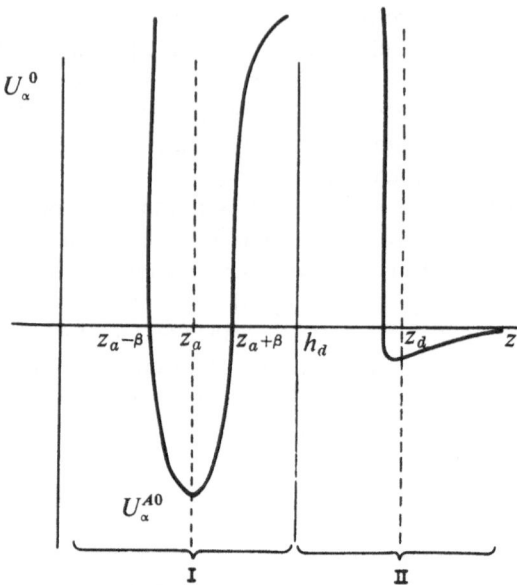

Fig. 5. The initial potential energy of adsorption
(at $q^m = 0$ and vanishing coverage of water and ionic
species) as a function of distance.

homogeneous metallic surface and limit our consideration to the case
of only one specifically adsorbed species.

3.6. General Expression of the Adsorption Isotherm

Whenever a particle is adsorbed in Region I and remains in it for
some short but finite time, this particle looses at least one of its degree
of translational freedom, which under these conditions is replaced by
vibrational motion normal to the surface.*. It means for an isolated
particle in close contact with the metal that the potential energy in the
direction z exhibits a deep or narrow minimum (Fig. 5). The depth
of the potential energy well U_α^{A0} represents the so-called initial heat of
adsorption measured at $q^m = 0$ and at zero coverage of the solvent
and the ions. The value of U_α^{A0} is of course determined by the nature
of the adsorptive forces between the adsorbate and adsorbent and
includes a coulombic self-image term so that $U_\alpha^{A0} = U_\alpha^{A00} + \frac{1}{2}(e^2{}_\alpha/2z_d)$.

---

* It occurs sometimes that the adsorption of α may be associated with restricted
rotation.

According to the form of the potential energy well, the particle α, once specifically adsorbed, is confined to lie within a very short distance from the minimum. This implies also that the centers of all adsorbed α, under normal conditions of adsorption of solvent and α, are essentially located in the very neighborhood of the plane z_a, so that the two-dimensional singlet distribution function ρ_α^A can be defined as follows [4]:

$$\rho_\alpha^A = \int_{z_a-\beta}^{z_a+\beta} \rho_\alpha^{(1)}(z)\, dz \tag{107}$$

where β, a small positive quantity, confirms the conditions $\beta \ll z_a$ and $\beta \ll h_d - z_a$.

Let us now reconsider equation (39). At the exchange equilibrium with the bulk phase, one obtains, according to (39) and Section 2, equation (25), that

$$\rho_\alpha^{(1)}(\{1\}) = a'_\alpha \exp[\mu'^0 - \mu^0(\{1\})]$$

$$\times \exp\left[-\frac{1}{kT}\left\{U_\alpha^{A0}(\{1\}) + A - kT \sum_{(\mathbf{n}-\alpha)\geqslant 1} \frac{1}{[\mathbf{n} - \alpha]!}\right.\right.$$

$$\left.\left. \times \int S\{\mathbf{n}\}\rho_2^{(1)}(\{2\})\rho_3^{(1)}(\{3\}) \cdots \rho_n^{(1)}(\{n_v\})\, d\{\mathbf{n} - 1\}\right\}\right] \tag{108}*$$

where a'_α is the bulk activity of α, the value of A is related to the boundary conditions of the electrostatic problem [Section 1, equations (11) to (14)], and where μ'^0 and $\mu_\alpha^0(\{1\})$ are, respectively, the standard-state chemical potential of species α in the bulk and at site (1). These standard-state functions may be expressed, as usual, in terms of the nonconfigurational part of the canonical partition function Q_T, Q_R, and Q_V, where, respectively, T, R, V refer to translational, rotational, and vibrational degrees of freedom. Hence,

$$\mu_\alpha'^0 - \mu_\alpha^0(\{1\}) = kT \ln\left(\frac{Q_T(\{1\})\, Q_R(\{1\})\, Q_V(\{1\})}{Q'_T Q'_R Q'_V}\right) \tag{109}$$

Each term of the irreducible cluster sum $S\{\mathbf{n}\}$ in (108) represents a product of cluster functions, as defined in (41). The characteristic

---

* Henceforth, it is assumed that integration over the entire range of internal coordinates is performed (e.g., water dipole moments).

manoeuver of the cluster theory in ionic solution is based on the following reordering of each of these functions;

$$
\begin{aligned}
f_{ij}(r_{ij}) &= \exp\left(-\frac{u_{ij}^*(r_{ij}) + u_{ij}^c(r_{ij})}{kT}\right) - 1 \\
&= \left[\exp\left(-\frac{u_{ji}^*(r_{ij})}{kT}\right) - 1\right] + \exp\left(-\frac{u_{ij}^*(r_{ij})}{kT}\right)\left[\exp\left(-\frac{u_{ij}^c(r_{ij})}{kT}\right) - 1\right] \\
&= f_{ij}^*(r_{ij}) + \exp\left(-\frac{u_{ij}^*(r_{ij})}{kT}\right) f_{ij}^c(r_{ij})
\end{aligned}
\tag{110}
$$

where $f_{ij}^*(r_{ij})$ defines a cluster bond of short-range nature; $u_{ij}^c(r_{ij})$ and $f_{ij}^c(r_{ij})$ represent, respectively, the coulomb contribution to the pairwise component potential and the resultant cluster function of coulombic nature.

In order to proceed further, it is convenient to split the sum of cluster integrals in (108) into two distinct parts. The terms included in the first of these parts correspond exclusively to graph interconnecting α at (1), which is now the only representative ion, with n_w solvent molecules. The second part will contain the remaining terms of (108). In view of the foregoing considerations, the sum in the exponent of (108) is given schematically by

$$
\sum_{(n-\alpha)\geqslant 1} = \sum_{n_w\geqslant 1} + \sum_{(n-\alpha)\geqslant 1}'
\tag{111}
$$

The second summation of (111) is performed over the composition sets $(n - \alpha)$ to the exclusion of (n_w) molecules. We may now introduce the following function:

$$
\Lambda(\rho_\alpha^{(1)}) \equiv \exp \sum_{(n-\alpha)\geqslant 1}' \frac{1}{[n - \alpha]!}
$$
$$
\times \int S\{n\}\rho_2^{(1)}(\{2\})\, \rho_3^{(1)}(\{3\}) \cdots \rho_{n_v}^{(1)}(\{n_v\})\, d\{n - 1\}
\tag{112}
$$

As mentioned earlier, in connection with equation (63), the treatment of the cluster integral (112) may be considerably simplified in assuming a statistical decoupling between the particles in the diffuse layer and the adions [4].* The reason for expecting such behavior lies in the main

---

* The effect of the diffuse layer ion distribution on the electrostatic interaction in the inner layer has been evidenced in some more detail by Ershler [6] and recently by Levine et al. [84].

features of our model of double layer as described in Fig. 1. These facts are (1) the marked spatial separation between the ions adsorbed in phases I and II, (2) the difference of dielectric constant between phase II and phase I, corresponding as indicated in Section 1, equation (10) to a virtual shift of the diffuse layer on a distance z_v away from the IHP, and (3) the greater mobility of the ions in Region II, which determine, during the lifetime of one adsorbed species at z_a, an averaging of the charge distribution over parallel planes to the IHP.

In the light of these observations, the interaction between α at (1) and any ion in the diffuse layer may be reasonably described by means of the first term of Mayer's linearized cluster expression; hence, according to Section 1, equations (3d) and (10), by a simple, pairwise coulombic bond of the form

$$f_{\alpha\gamma}^c(r_{\alpha\gamma}) = -\frac{e_\alpha e_\gamma}{kT\epsilon_0}\gamma(r_{\alpha\gamma}) = -\frac{e_\alpha}{kT}\psi_{\gamma,\mathrm{I}}^D \qquad (113)$$

The effect of graphs of higher order [in $\gamma(r)$] is presumed to be small and, as usual, the influence of the solvent is implied in ϵ_0 and ϵ_1. It is now possible to perform the integration in (112) separately, over Regions I and II, thus, on account of Section 1, equation (11), one has

$$\Lambda(\rho_\alpha^{(1)}) = \Lambda_\alpha(\rho_\alpha{}^A)\exp\left(-\frac{\bar\psi(z_a)}{kT}\right) \qquad (114)$$

with

$$\Lambda_\alpha(\rho_\alpha{}^A) \equiv \exp\sum_{(n-\alpha)\geqslant 1}' \frac{1}{[\mathbf{n}-\alpha]!}$$

$$\times\int_{\mathrm{I}} S\{\mathbf{n}\}\rho_2^{(A)}(\{2\})\,\rho_3^{(A)}(\{3\})\cdots\rho_{n_\nu}^{(A)}(\{n_\nu\})\,d\mathbf{s}_{12}\,d\mathbf{s}_{13}\cdots d\mathbf{s}_{1n_\nu})$$

$$+\frac{1}{kT}\int_{\mathrm{I}} d\mathbf{s}_{12}\,\frac{e_\alpha^2}{\epsilon_0}\gamma^A(s_{12})\,\rho_\alpha^{(A)}(\{2\}) \qquad (115)$$

In the exponent of (115), integration along z has been performed according to (107). Owing to equation (114) and (115), the expression of the adsorption isotherm can next be given under the form

$$\rho_\alpha{}^A = a_\alpha' l_\alpha K\,\exp\left(-\frac{e_\alpha\bar\psi(z_a)}{kT}\right)\cdot\Lambda_\alpha(\rho_\alpha{}^A) \qquad (116)$$

where the specific length l_α and constant K_α are defined as follows:

$$l_\alpha K_\alpha \equiv \int_{z_{a-\beta}}^{z_{a+\beta}} \exp \left\{ - \frac{1}{kT} [\mu_\alpha^0(\{1\}) - \mu'^0 + U_\alpha^{0A}(\{1\})] \right\} dz$$

$$\times \exp \sum_{n_w \geqslant 1} \frac{1}{n_w!} \int_{\mathrm{I}} S\{\mathbf{n}\} \rho_w^A(\{2\}) \, \rho_w^A(\{3\}) \ldots \rho_w^A(\{n_w\}) \, d\mathbf{s}_{12} \, d\mathbf{s}_{13} \ldots d\mathbf{s}_{1n_w}$$

$$(117)$$

Each graph of the cluster sum $S\{\mathbf{n}\}$ in (117) is rooted at the position of α at (1) and contains n_w water dipoles in addition to ion α. Thus, we note that each term of the cluster sum is a product of $f_{\alpha w}$ and f_{ww} bonds, the subscript w indicating the solvent molecule. Owing to the large occupation of the surface by the solvent molecules, a primary role may be granted to size effects. Consequently, the electrostatic interaction between ion α and dipole w can be regarded as a perturbation to such size effects. Therefore, we assume to keep in each cluster sum besides the unperturbed graphs (of $f_{\alpha w}^*$ and f_{ww}^* bonds) only those graphs linearly perturbed by bond $f_{\alpha w}^c$. According to this procedure, the cluster sum may be split into two terms; (1) a sum of unperturbed graphs $S^0\{\mathbf{n}\}$ and (2) a sum of linear perturbed graphs obtained through replacement in each term of $S^0\{\mathbf{n}\}$ of one single bond $f_{\alpha w}^*$ by $\exp[(-u_{\alpha w}^*(s_{12})/kT) f_{\alpha w}^c(s_{12})]$. It should be noted that this perturbing bond may act between α and n_w distinct vertices, so that (117) finally yields, in an expected way,

$$l_\alpha K_\alpha \equiv l_\alpha H_\alpha \exp \left[-\frac{1}{kT} \int_{\mathrm{I}} d\mathbf{s}_{12} u_{\alpha w}^c(s_{12}) \, \rho_w^A(s_{12}) \, g_{\alpha w}^{0(2)}(s_{12}) \right] \qquad (118)$$

in which $f_{\alpha w}^c$ has been linearized and where

$$l_\alpha H_\alpha \equiv \int_{z_{a-\beta}}^{z_{a+\beta}} \exp \left[-\frac{1}{kT} (\mu_\alpha^0 \{1\} - \mu_\alpha'^0 + U_\alpha^{A0}(\{1\})) \right] dz$$

$$\times \exp \left[\sum_{n_w \geqslant 1} \frac{1}{n_w!} \int_{\mathrm{I}} S^0\{\mathbf{n}\} \rho_w^A(\{2\}) \, \rho_w^A(\{3\}) \cdots \rho_w^A(\{n_w\}) \, d\mathbf{s}_{12} \, d\mathbf{s}_{13} \cdots d\mathbf{s}_{n_w} \right]$$

$$(119)$$

In (118), $u_{\alpha w}^c$ is the electrostatic interaction energy between α and a water dipole in its state of average configuration at the surface. If $g_{\alpha w}^{0(2)}(s_{12})$, the unperturbed pair correlation function, is assumed to be unity, the adsorbed water dipoles constitute a double layer of charge smeared out over two parallel planes to the electrode surface* and

* A cavity field effect results from condition $g_{\alpha w}^{0(2)} = 0$ at the adsorption place of α.

situated on each side of z_a. Furthermore, if $\langle \bar{\mu} \rangle$ represents the component along z of the average dipole moment, the application of simple electrostatics suffice to provide an approximate solution of (118) of the form

$$l_\alpha K_\alpha = l_\alpha H_\alpha \exp \left[-\frac{4\pi e_\alpha}{kT\epsilon_0} \frac{h_d - z_a}{h_d} \rho_w{}^A \langle \bar{\mu} \rangle \right]$$

$$= l_\alpha H_\alpha \exp \left[-\frac{e_\alpha}{kT} \frac{h_d - z_a}{h_d} \bar{\chi} \right] \tag{120}$$

The potential $\bar{\chi}$ is known as the surface potential. Some detailed treatments of the potential $\bar{\chi}$ have been given recently [71,73]. The value of ϵ_0 has been derived by different authors and in most cases is about 4 to 6 [71,74].

In collecting the results obtained above we are now in a position to write at vanishing coverage that

$$\lim_{\rho_\alpha^A \to 0} \rho_\alpha{}^A = a' l_\alpha H_\alpha \exp \left[-\frac{e_\alpha}{kT} \left\{ \left(\frac{h_d - z_a}{h_d} \right) \bar{\chi}(q_a = 0) + \bar{\psi}_{q_a=0}(z_a) \right\} \right] \tag{121}$$

3.7. The Mobile Monolayer Adsorption

3.7.1. *Description of the Model of Mobile Adsorption.* Let us consider an adsorbed monolayer on a plane, homogeneous metallic surface. Under these conditions, it is known that the range of the electrostatic forces $e_\alpha \gamma(s_{12})/\epsilon_0$ acting between pairs of adions is considerably smaller than $1/r_{12}$. The most general expression of $\gamma^A(r_{12})$ has been derived in Section 1, equation (7b) and was shown to be somewhat involved. However, in the case of a single-image model ($\delta = 0$), it is readily deduced that for $s_{12} > a \geqslant 2z_a$,* one has

$$\gamma^A(s_{12}) = \frac{1}{s_{12}} \left[1 - \frac{1}{[1 + 4z_a{}^2/s_{12}{}^2]^{1/2}} \right] \simeq \frac{2z_a{}^2}{s_{12}{}^3} + O(s_{12}^{-5}) \tag{122}$$

It is easily inferred from Section 1, equation (7b) that multiple reflection into the planes at $z = 0$ and $z = h_d$ has an even more

---

* The radius of the hard-core exclusion sphere a should not be expected to be necessarily equal to $2z_a$. On the basis of some quantum mechanical correction [75,76], and assuming the validity of representing an ion by a point charge at its center, the correction to the classical value of image forces is less than 10 % for z_a of about 1 to 2 Å. This correction is not further considered here.

decreasing effect on the range of this force. Let us introduce now the approximate mean distance between two neighboring adsorbed ions, $s_{12} = s_0$;

$$\rho_\alpha{}^A \pi \left(\frac{s_0}{2}\right)^2 \simeq 1 \qquad (123)$$

Considering the case of a specifically adsorbed charge density of $10 \ \mu\text{Cb/cm}^2$, $\epsilon_0 \simeq 6$ and $z_a \simeq 2$ Å, we compute easily that $s_0 = 14$ Å and

$$\frac{e_\alpha{}^2}{\epsilon_0} \gamma^A(s_0) \leqslant 0.3 \ kT \qquad (124)$$

Assuming that in an ordered arrangement extended as far as the first neighbors, a coordination of α with six adions may occur and that in the extreme and rather improbable case of a complete hexagonal array extended over the entire IHP, with a lattice parameter s_0, the coordination number is about 11, we see that the total coulombic interaction does not exceed respectively $1.8 \ kT$ and $3.3 \ kT$. Accordingly, some stress perhaps should be laid on the fact that in the range of coverages $\leqslant 10 \ \mu\text{Cb/cm}^2$ thermal motion cannot be neglected in the computation of distribution functions at the IHP. At least, for such coverages, the foregoing argument leads to questioning the applicability of a model of artificial hexagonal rigid and two-dimensional lattice at a liquid metal surface if no thermal motion is allowed to smooth the distribution function.* Let us now appreciate the situation at very low coverage, less than $1 \ \mu\text{Cb/cm}^2$, and therefore introduce two distinct parameters P_1 and P_2. For the first parameter we write $P_1 = \rho_\alpha{}^A/\rho_{\alpha\text{cp}}^A = s_{\text{cp}}^2/s_0^2$, where s_0 is the mean distance defined above and s_{cp} is the equidistance at close packing at the IHP, so that $\rho_{\alpha\text{cp}}^A \pi (s_{\text{cp}}/2)^2 \simeq 1$. The second parameter P_2 is $P_2 = (e_\alpha{}^2/\epsilon_0 kT) \gamma^A(s_0)$. In the bulk phase under Debye–Hückel's conditions, P_1 is proportional to the ionic strength and P_2 to the cubic root of the ionic concentration [cf. equation (87)]. However, at the surface of adsorption one has $P_1 = \rho_\alpha{}^A/\rho_{\alpha\text{cp}}^A$ and $P_2 \simeq (\rho_\alpha{}^A/\rho_{\alpha\text{cp}}^A)^{3/2}$. Owing to the shortened range of action of the coulombic forces, it appears, thus, that unlike the Debye–Hückel type of behavior with decreasing concentration $\rho_\alpha{}^A$, the electrostatic contribution resulting from discreteness of charges decreases faster than do the effects of finite ionic size. The previous consideration thus allows, at the limit of low coverages to retain for the computation

---

* The situation for hexagonal lattice at large coverage is discussed in Section 3.8.

of $\Lambda_\alpha(\rho_\alpha{}^A)$, a nonperturbed form of a singlet potential of average force determined in other respect for a two-dimensional fluid of noncharged particles [4]. However, at larger coverage, such approximation becomes insufficient because the discreteness of charge effect may contribute to the potential of average force by a term of the order of kT. Therefore, we shall expand the irreducible cluster sum $S\{\mathbf{n}\}$ in $\Lambda_\alpha(\rho_\alpha{}^A)$ as follows, starting from the noncoulombic perturbed terms $S^0\{\mathbf{n}\}$, hence,

$$S\{\mathbf{n}\} = \sum_{n_\lambda > i,j} \prod \left\{ f_{\alpha\alpha}^*(s_{ij}) + \exp\left(-\frac{u_{\alpha\alpha}^*(s_{ij})}{kT}\right) f_{\alpha\alpha}^c(s_{ij}) \right\}$$

$$= S^0\{\mathbf{n}\} + \frac{n_\alpha!}{2!(n_\alpha - 2)!} f_{\alpha\alpha}^c(s_{12}) \exp\left[-\frac{u_{\alpha\alpha}^*}{kT}(s_{12})\right] C\{\mathbf{n} - 1, 2\}$$

$$+ \frac{n_\alpha!}{3!(n_\alpha - 3)!} [3 f_{\alpha\alpha}^c(s_{12}) f_{\alpha\alpha}^c(s_{23}) + f_{\alpha\alpha}^c(s_{12}) f_{\alpha\alpha}^c(s_{23}) f_{\alpha\alpha}^c(s_{31})]$$

$$\exp\left[-\frac{u_{\alpha\alpha}^*(s_{12}) + u_{\alpha\alpha}^*(s_{31}) + u_{\alpha\alpha}^*(s_{31})}{kT}\right] C\{\mathbf{n} - 1, 2, 3\} \qquad (125)$$

The first term $S^0\{\mathbf{n}\}$ represents a sum of products of unperturbed cluster functions $f_{\alpha\alpha}^*$. In the second term one of these unperturbed bonds has been replaced by $f_{\alpha\alpha}^c \exp -u_{\alpha\alpha}^*/kT$. The binomial coefficient is the number of ways of isolating a distinct pair of vertices out of a collection of n_α entities. The electrostatic perturbation can act between any of these pairs of vertices, which in each case on account of translational invariance in the plane z_a may be renumbered so that the coulombic bond acts between 1 and 2. The third term would correspond to the case of two successive perturbed bonds and the fourth to a closed cycle of perturbed bonds on the skeleton of vertices 1, 2, and 3. The quantities $C\{\mathbf{n} - 1, 2\}$ and $C\{\mathbf{n} - 1, 2, 3\}$ represent a sum of products of unperturbed bonds linked respectively with 1, 2 and 1, 2, 3. These sums have already been introduced in connection with (40).

It is also worth emphasizing at this point that the cluster sum (125) will be taken exclusively on the set of n_α adions. As usual, the effect of the solvent is implied in the definition of the value of ϵ_0 and of $f_{\alpha\alpha}^*$ (this will have to be remembered later for the choice of $u_{\alpha\alpha}^*$).

In the present model of mobile adsorption, the singlet function $\rho_\alpha{}^A$ in (115) may be taken outside the integrals so that finally, according to (125), we obtain for the unperturbed part,

$$\Lambda^0(\rho_\alpha{}^A) \equiv \exp\left[\sum_{(n_\alpha - 1) \geqslant 1} \frac{(\rho_\alpha{}^A)^{n_\alpha - 1}}{(n_\alpha - 1)!} \int_{\mathrm{I}} S^0\{n_\alpha\} \, d\mathbf{s}_{12} \, d\mathbf{s}_{13} \cdots d\mathbf{s}_{1n_\alpha}\right] \qquad (126a)$$

On application of a first-density derivative, as suggested by Buff and Stillinger [4] and according to relations (16), (19) and (43), one is led finally to

$$
\begin{aligned}
\Lambda_\alpha(\rho_\alpha{}^A) = \Lambda_\alpha{}^0(\rho_\alpha{}^A) \exp \Bigg[& \int_{\mathrm{I}} d\mathbf{s}_{12} f_{\alpha\alpha}^c(s_{12}) \\
& \times \frac{1}{2} \frac{\partial}{\partial \rho_\alpha{}^A} (\rho_\alpha{}^A)^2 \cdot g_{\alpha\alpha}^{0(2)}(s_{12}) \left\{ 1 + \rho_\alpha{}^A \int_{\mathrm{I}} d\mathbf{s}_{13} (f_{\alpha\alpha}^c(s_{23})) \right. \\
& \left. + \frac{1}{3} f_{\alpha\alpha}^c(s_{13}) f_{\alpha\alpha}^c(s_{23})) g_{\alpha\alpha}^{0(2)}(s_{23}) \cdot g_{\alpha\alpha}^{0(2)}(s_{13}) + \cdots \right\} \\
& + \frac{\rho_\alpha{}^A}{\epsilon_0 kT} \int_{\mathrm{I}} d\mathbf{s}_{12} e_\alpha{}^2 \gamma^A(s_{12}) \Bigg]
\end{aligned}
\tag{126b}
$$

Since the $C\{\mathbf{n}\}$ are unperturbed cluster sums, it is clear that $g_{\alpha\alpha}^{0(2)}$ corresponds to the pair correlation function in a two-dimensional fluid where particles interact just through short-range $f_{\alpha\alpha}^*$ bonds.

Our plan is to evaluate first $\Lambda_\alpha{}^0(\rho_\alpha{}^A)$ and to find next a reasonable model of radial distribution.

As in (81), we assume for convenience that the short-range forces between pairs of adions α can be described through a model of hard disks (the two-dimensional counterpart of the hard-sphere model) with a common radius a. As stated previously, in view of the effect of water on the value of $f_{\alpha\alpha}^*$, the radius a would probably have to differ from the ionic diameter, $u_{\alpha\alpha}^*$ behaving, by definition, as a potential of mean force for a set of two discharged α particles in interaction with the solvent. The conditions for the hard-sphere model are

$$
\begin{aligned}
u_{\alpha\alpha}^*(s_{12}) = 0 && f_{\alpha\alpha}^*(s_{12}) = 0 && \text{for} \quad s_{12} \geqslant a \\
u_{\alpha\alpha}^*(s_{12}) = \infty && f_{\alpha\alpha}^*(s_{12}) = -1 && \text{for} \quad s_{12} \leqslant a
\end{aligned}
\tag{127}
$$

We may now conclude from (116), (120), and (126a) that

$$
\rho_\alpha{}^A = a' \Lambda_\alpha{}^0(\rho_\alpha{}^A)
\tag{128}
$$

is just the adsorption isotherm for a fluid of hard spheres. In this case the Gibbs' equation yields

$$
\begin{aligned}
\left(\frac{\partial \Pi}{\partial \rho_\alpha{}^A} \right)_{T,\mathfrak{p}} &= \rho_\alpha{}^A \left(\frac{\partial \mu_\alpha}{\partial \rho_\alpha{}^A} \right)_{T,\mathfrak{p}} = kT \left(\frac{\partial \ln a_\alpha'}{\partial \ln \rho_\alpha{}^A} \right)_{T,\mathfrak{p}} \\
&= kT \left(1 - \frac{\partial \ln \Lambda_\alpha{}^0(\rho_\alpha{}^A)}{\partial \ln \rho_\alpha{}^A} \right)_{T,\mathfrak{p}}
\end{aligned}
\tag{129}
$$

where Π is the spreading pressure. In the usual procedure, (129) is combined to the equation of state

$$\left(\frac{\Pi}{\rho_\alpha{}^A kT} - 1\right) = \rho_\alpha{}^A \frac{\pi a^2}{2} g_{\alpha\alpha}^{0(2)}(s_{12} = a) \tag{130}$$

The equation of state for a hard-disk fluid has been recently evaluated by different authors. Helfand, Frish, and Lebowitz [77] obtained

$$g_{\alpha\alpha}^{0(2)}(a) = \frac{1 - y/2}{(1 - y)^2} \tag{131a}$$

where $y = \rho_\alpha{}^A(\pi a^2/4)$. The virial relation derived by Ree and Hoover [78], however less convenient in its expression, is probably more accurate. It gives

$$g_{\alpha\alpha}^{0(2)}(a) = \frac{1 - (0.1967)(2y) + (0.006519)(2y)^2}{[1 - 0.489351(2y)]^2} \tag{131b}$$

Comparing (131a) and (131b), it appears that these results, although obtained by different methods, are very close to each other. The integration of (129), combined with (130) and one of the two values of $g_{\alpha\alpha}^{0(2)}(a)$, yields

$$\ln \Lambda^0(\rho_\alpha{}^A) = A \ln (1 - 2Cy) - \frac{2Cy}{(1 - 2Cy)^2} \{B + 2DCy + E(2Cy)^2\} \tag{132}$$

Parsons [79] and Buff and Stillinger [4] have shown that with (131a), $A = 1$, $B = 3$, $C = 1/2$, $D = -2$, $E = 0$. Parsons [79] has also compared (132) to other isotherms. From (131b) it may be inferred that $A = 0.714$, $B = 3.286$, $C = 0.4893$, $D = -2.167$, and $E = 0.111$.

It is finally worth mentioning that at close packing of the disks, y is equal to 0.904. Thus, the relative degree of coverage θ_c corresponds to $\theta_c = y/0.904$.

Any attempt to proceed further in solving (126) requires the determination of $g_{\alpha\alpha}^{0(2)}(s_{12})$ and of the corresponding potential of mean force $W_{\alpha\alpha}^{0(2)}$. Whereas the three-dimensional problem has been recently successfully tackled [80,81], there is at present no convenient expression available for the radial distribution in a fluid of hard disks. There exists only at low coverage ($y \ll 0.1$) an easy way to determine $g_{\alpha\alpha}^{0(2)}$ within an accuracy of the second virial coefficient of the equation of state [42]. However, we may expect that the correlation function $g_{\alpha\alpha}^{0(2)}(s_{12})$ corresponding to the hard-sphere model vanishes for $s_{12} \leqslant a$

and oscillates around unity for $s_{12} \geqslant a$ (Fig. 6). According to such behavior, it next will be assumed that in (126b) we have

$$g_{\alpha\alpha}^{0(2)}(s_{12}) \left\{ 1 + \rho_\alpha{}^A \int_I d\mathbf{s}_{13} f_{\alpha\alpha}^c(s_{23}) g_{\alpha\alpha}^{0(2)}(s_{23}) g_{\alpha\alpha}^{0(2)}(s_{31}) + \cdots \right\}$$

$$= 0 \qquad\qquad \text{for} \quad s_{12} \leqslant a \qquad\qquad (133)$$

$$\simeq \{1 + \Gamma(s_0)[\eta(s') - \eta(s'')]\} \quad \text{for} \quad s_{12} \geqslant a$$

where $\eta(s')$ and $\eta(s'')$ are the Heaviside functions and $\Gamma(s_0)$ is the magnitude of the steps at s' and s'', situated on either side of $s_{12} = s_0$.

Before going further, we wish to stress here some of the basic implications of the model given in (133). This model concerns some correlation between two adions (without direct coulombic connection $\gamma^A(s_{12})$) situated respectively at (1) and (2). From numerical calculations in the model of hard spheres it is known that the superposition of unperturbed $g_{\alpha\alpha}^{0(2)}(s_{23}) g_{\alpha\alpha}^{0(2)}(s_{13})$... give rise in the immediate vicinity of α at (1) to some slightly ordered arrangement. At larger distance the distribution becomes uniform [42]. However, the effect of repulsion in $f_{\alpha\alpha}^c(s_{23})$ will now attempt to move away these neighbors of α at (1) so that the probability of finding an adion at (2) is also decreased near (1) and assumed to be enhanced at s_{12} approximately equal to the mean distance s_0. It is further granted that the interaction of α at (1) with its adion atmosphere at s_0 can be characterized through a number N, a pseudo-coordination number not to be confused with the real coordination number related to the correct radial distribution $g_{\alpha\alpha}^{(2)}$; thus by definition

$$N = \rho_\alpha{}^A \lim_{\substack{s' \to s_0 \\ s'' \to s_0}} \int_0^{2\pi} (s'' - s') \cdot s_0 \Gamma(s_0) \, d\theta \qquad\qquad (134)$$

The resulting value of the radial function (133) is illustrated in Fig. 6.

Upon introduction of equations (133) and (134) into (126) and after linearization of $f_{\alpha\alpha}^c(s_{12})$ according to (124) one readily finds

$$-kT \ln \frac{\Lambda_\alpha(\rho_\alpha{}^A)}{\Lambda_\alpha^0(\rho_\alpha{}^A)} = -\rho_\alpha{}^A \int_0^a d\mathbf{s}_{12} \frac{e_\alpha{}^2}{\epsilon_0} \gamma^A(s_{12})$$

$$+ N \frac{e_\alpha{}^2}{\epsilon_0} \gamma^A(s_{12}) + \lim_{\substack{s' \to s_0 \\ s'' \to s_0}} \frac{(\rho_\alpha{}^A)^2}{2} \int \frac{e_\alpha{}^2}{\epsilon_0} \gamma^A(s_{12})$$

$$\times \frac{\partial}{\partial \rho_\alpha{}^A} (\Gamma(s_0)(\eta(s') - \eta(s''))) \, d\mathbf{s}_{12} \qquad\qquad (135)$$

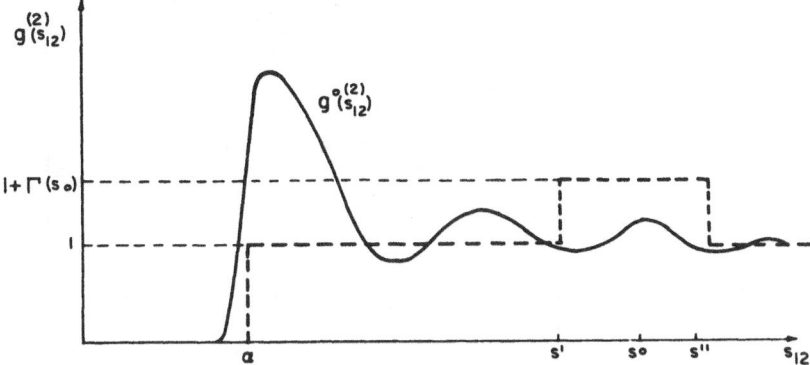

Fig. 6. The model of a pair correlation function; $g^{0(2)}(s_{12})$ is the unperturbed function for a system of noncharged hard disks.

3.7.2. *The Cavity Field.* It is not difficult to recognize in the first term of the right-hand side of equation (135) the average electrostatic potential produced by the specifically adsorbed charges that would lie, on the average, in a circular cavity of radius a created by particle α when adsorbed at (1). Thus, it is apparent that such a term is just a correction to the average electrostatic potential or macroscopic potential $\bar{\psi}(z_a)$, since the adsorption of a particle of finite size has prevented the uniform distribution to spread over the entire surface. Hence, we may write

$$\bar{\psi}(z_a) - q_a \frac{2\pi}{\epsilon_0} \int_0^a s_{12}\gamma^A(s_{12})\,ds_{12} = \bar{\psi}_\alpha^A(\text{cav}) \qquad (136)$$

The cavity potential so defined has been evaluated by Levine et al. [69]* and in a slightly different way by a number of authors [67,68]. In the case of multiple reflection into two perfect conducting plates at $z = h_d$ and $z = 0$ ($\epsilon_0 \ll \epsilon_1$) the parameter δ in Section 1, equation (17b) is unity and the problem consists in determining the field in a small circular cavity between two conducting plates at zero potential, distant by h_d. For $2a > h_d$, the argument presented by Levine et al. [69] is

---

* Levine et al. [69] have given to this correction a different meaning in extending the cavity over the distance $s_{12} = s_0$. Their model thus includes size and atmosphere effects.

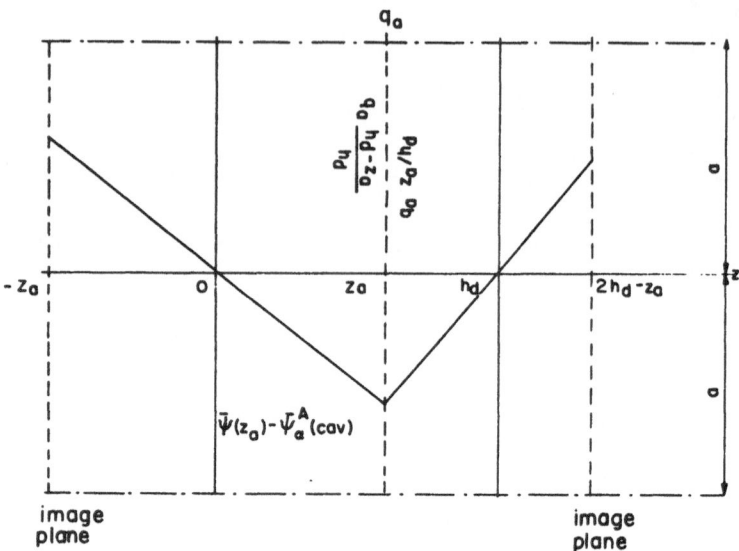

Fig. 7. Calculation of the electrostatic potential drop in the inner layer for a charge density q_a on a circular surface of radius a between two conducting plates (cf. Levine *et al.* [69]).

adopted here. In Fig. 7 we have split the charges q_a on the IHP into $q_a z_a / h_d$ and $q_a(h_d - z_a)/h_d$, which are imagined to lay, respectively, on the side of the IHP facing the plane at h_d and the electrode. The potential drop between the electrode and z_a, as well as between z_a and h_d, is therefore $4\pi q_a (h_d - z_a/h_d) \cdot z_a$.

Consequently, the cavity potential under these conditions is

$$\bar{\psi}(z_a) - 4\pi q_a z_a \cdot \frac{h_d - z_a}{h_d}\, (1 - \beta(h_d/a)) = \bar{\psi}_\alpha{}^A(\mathrm{cav}) \qquad (137)$$

where $(1 - \beta(h_d/a))$ accounts for the edge effects at $s_{12} = a$ and has been evaluated by Levine [69]. The value of the coefficient $\beta(h_d/a)$ is given in Fig. 8 in function of the ratio h_d/a [82].

Whenever $\epsilon_1 \simeq \epsilon_0$, $\delta = 0$, one finds that

$$\bar{\psi}(z_a) - q_a \frac{2\pi}{\epsilon_0} \left[a + 2z_a - a \sqrt{1 + \left(\frac{2z_a}{a}\right)^2}\right] = \bar{\psi}_\alpha{}^A(\mathrm{cav}) \qquad (138)$$

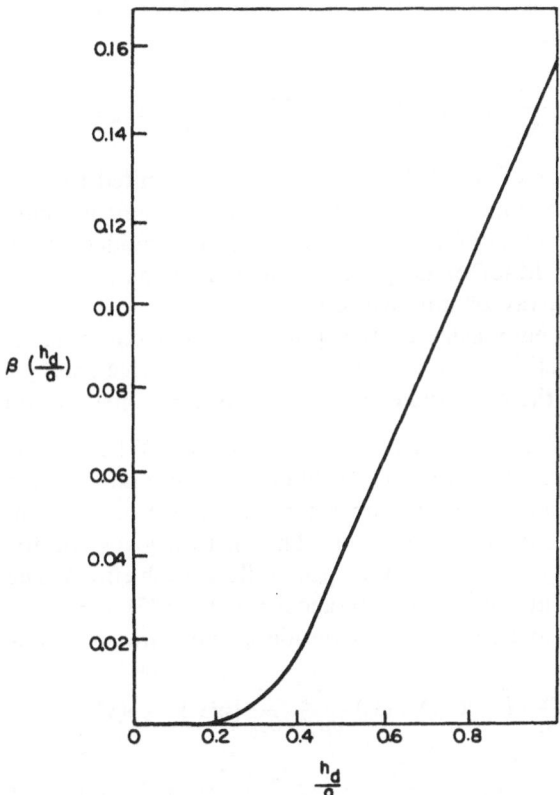

Fig. 8. The coefficient $\beta(h_d/a)$ of Levine *et al.* [69] as plotted in Krylov [82].

3.7.3. *The Coordination Field.*

In order to bring out the second term of (135) we recall that the model considered in (133) assumes that the pair correlation function of unity is increased by $\Gamma(s_0)$ at the mean distance s_0. A corresponding amount of N ions is then superposed near s_0 upon the uniform distribution in the plane. A great number of authors have suggested substituting for the whole layer a regular hexagonal array of ions whose spacing is given by s_0 [65–68]. As already emphasized, in such a model no provision is made for thermal motion.

In the case where $\delta = 1$, we refer the reader to the work of Krylov, Kiryanov, and Levich [67] and of Barlow and MacDonald [68]. This problem is reconsidered in Section 3.8 for the micropotential.

In the case where $\delta = 0$, one finds readily on account of (123) and assuming (122)

$$\frac{e_\alpha^2}{\epsilon_0} N \gamma^A(s_0) \simeq \frac{e_\alpha^2}{\epsilon_0} N \frac{2z_a^2}{s_0^3} \simeq \frac{e_\alpha^2}{4\epsilon_0} N z_a^2 (\pi \rho_\alpha^A)^{3/2} \tag{139}$$

The accuracy of the following treatment is limited to the ideal dipole contribution considered in (122). A more rigorous computation is beyond the limits of accuracy allowed by our model. Mignolet [83] and Barlow and MacDonald [68] have calculated the interaction energy of an infinite array of nonideal dipoles.

Let us emphasize at this point that the coordination field here defined is fictitious. In fact it yields the excess free energy of the adion system, not the potential energy, and N is a parameter of the treatment.

3.7.4. *The Radial Redistribution Work.* Whenever we introduce in our system at position (1) a differential amount $\delta\rho_\alpha^A$ of the entity α, we do not only alter the local density but we modify simultaneously the radial distribution function. This in turn gives rise to some interaction energy. The last term of (135) reflects such effects. The integration of this term may be performed according to (123), (134), and the definition and properties of the Heaviside η and Dirac δ functions; thus,

$$\lim_{\substack{s' \to s_0 \\ s'' \to s_0}} \frac{(\rho_\alpha^A)^2}{2} \frac{e_\alpha^2}{\epsilon_0} \int \gamma^A(s_{12}) \, \Gamma(s_0) \frac{\partial s_0}{\partial \rho_\alpha^A} \frac{\partial}{\partial s_0} [\eta(s') - \eta(s'')] \, ds_{12}$$

$$= \lim_{\substack{s' \to s_0 \\ s'' \to s_0}} \frac{e_\alpha^2}{2\epsilon_0} \rho_\alpha^A \frac{N}{s_0(s'' - s')} \left(\frac{\partial s_0}{\partial \rho_\alpha^A} \right) \int \gamma^A(s_{12}) \{\delta(s'') - \delta(s')\} s_{12} \, ds_{12}$$

$$= -\frac{e_\alpha^2}{\epsilon_0} \frac{N}{4} \lim_{\substack{s' \to s_0 \\ s'' \to s_0}} \frac{s'' \gamma(s'') - s' \gamma(s')}{s'' - s'}$$

$$= -\frac{e_\alpha^2}{4\epsilon_0} N \left(\frac{\partial s \gamma^A(s)}{\partial s} \right)_{s \to s_0} \tag{140}$$

Upon introduction of the expressions (122) and (123) into (140) one finds for the case $\delta = 0$ that

$$(140) = \frac{e_\alpha^2}{4\epsilon_0} \frac{N}{2} (z_a)^2 (\pi \rho_\alpha^A)^{3/2} \tag{141}$$

It is worth pointing out that (140) results from our model (133) and not from rearrangement of the correct function $g_{\alpha\alpha}^{(2)}$. The redistribution work has been considered qualitatively by different authors [68,70] but generally ignored in the final formulation of an adsorption isotherm. In this respect it is noteworthy that the result (141) gives just half the value of (139).

    3.7.5. *The Final Form of the Adsorption Isotherm and its Application.* We may now collect the different expressions derived above in order to formulate the adsorption isotherm in its final version. We have

$$\frac{y}{(1 - 2Cy)^4} = a'_\lambda \frac{\pi a^2}{4} l_\lambda H_\alpha \exp \left\{ - \frac{2Cy}{(1 - 2Cy)^2} [B + 2DCy + E(2Cy)^2] \right\}$$

$$\times \exp - \frac{1}{kT} \left(e_\alpha \frac{h_d - z_a}{h_d} \bar{\chi} + W_\alpha^{c(1)} \right) \tag{142}$$

where the coulombic part of the singlet potential of mean force $W_\alpha^{c(1)}$ is given by

$$W^{c(1)} = e_\alpha \bar{\psi}(z_a) - \frac{q_a}{\epsilon_0} e_\alpha 2\pi \int_0^a s_{12} \, ds_{12} \gamma^A(s_{12})$$

$$+ \frac{Ne_\alpha}{4\epsilon_0} \left[3\gamma^A(s_0) - s_0 \left(\frac{\partial \gamma^A(s)}{\partial s} \right)_{s_0} \right] \tag{143}$$

In the case of $\epsilon_1 \gg \epsilon_0$, ($\delta = 1$), we finally get on account of (137) and, neglecting the potential drop in the diffuse layer,

$$W_\alpha^{c(1)} \simeq \frac{4\pi e_\alpha}{\epsilon_0} \left(\frac{h_d - z_a}{h_d} \right) \left[q^m h_d + q_a(h_d - z_a) + q_a \beta \left(\frac{h_d}{a} \right) z_a \right]$$

$$+ \frac{Ne_\alpha^2}{4\epsilon_0} \left[3\gamma^A(s_0) - s_0 \left(\frac{\partial \gamma^A(s)}{\partial s} \right)_{s_0} \right] \tag{144}$$

If $\epsilon_1 \simeq \epsilon_0$ and $\delta = 0$, we obtain for $a \simeq 2z_a$, according to (138), (139), and (141)

$$W_\alpha^{c(1)} \simeq \frac{4\pi e_\alpha}{\epsilon_0} \{ q^m (h_d - z_a) \} + \frac{4\pi e_\alpha^2}{\epsilon_0} \rho_\alpha^A \{ h_d - 1.6z_a \}$$

$$+ \frac{e_\alpha}{4\epsilon_0} (\pi \rho_\alpha^A)^{3/2} z_a \frac{3N}{2} \tag{145}$$

where again the average electrostatic potential drop in the diffuse layer is disregarded. The previous expression may be inserted into (142) and the equation of state (131a) assumed; thus,

$$\frac{y}{1-y} = a'_\alpha \pi z_a{}^2 l_\alpha H_\alpha \exp\left[-\frac{y}{(1-y)^2}(3-2y)\right]$$

$$\times \exp\left\{-\frac{e_\alpha}{kT}\left[\left(q^m\frac{4\pi}{\epsilon_0}+\frac{\bar{\chi}}{h_d}\right)(h_d-z_a)\right]\right\}$$

$$\times \exp\left\{-\frac{e_\alpha{}^2}{kT}\left[y\frac{4}{z_a{}^2}(h_d-1.6z_a)+y^{3/2}\frac{3N}{8z_a}\right]\right\} \qquad (146)$$

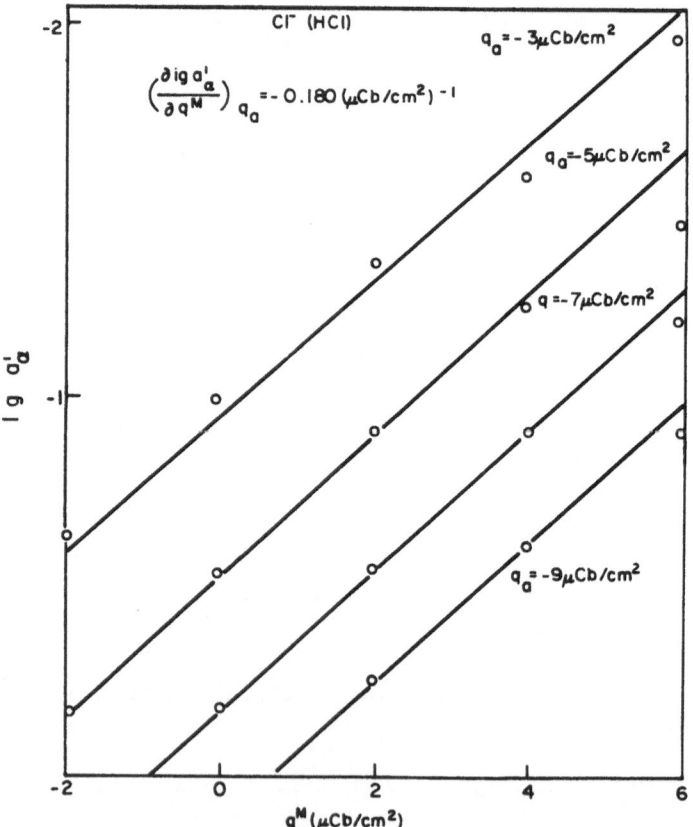

Fig. 9. Specific adsorption on Hg of Cl⁻ in HCl aqueous solution of mean activity a'_α. The logarithm (to base ten) of a'_α is plotted vs. the charge density of the electrode at constant specific adsorption q_a (cf. Wroblowa *et al.* [72]).

Whenever $\bar{\chi}$ is a linear function or is independent of q^m, y small, and $h_d \simeq 1.6z_a$, this expression yields back an isotherm suggested by Bockris *et al.* [71] on the basis of a somewhat arguable treatment. This isotherm, however, has been shown to be in good agreement with recent experimental results [72]. On the basis of this data, and assuming a linear change of $\bar{\chi}$ with q^m [71,73], we have made use of (146) for the adsorption of Cl^- on Hg at different values of q^m [85]. From the plots in Figs. 9 and 10, $N \simeq 9$ if $\epsilon_0 \simeq 6$, which is indeed a reasonable

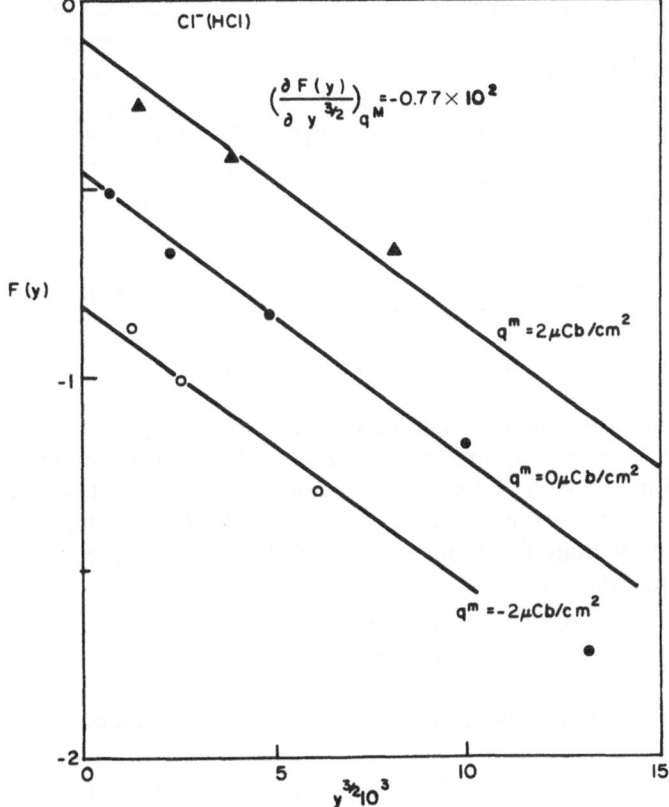

Fig. 10. The function

$$F(y) \equiv \log \frac{y}{(1-y)\,a'_\alpha} + \frac{y}{(1-y)^2} \frac{3-2y}{2.3}$$

plotted vs. $y^{3/2}$ according to the data recorded in Fig. 9 and in Wroblowa *et al.* [72].

figure. However, owing to the simplifying assumptions made above (diffuse layer) the experimental deduction should not be taken too literally.

3.8. The Discreteness of Charge Model and Lattice Adsorption

Owing to the possible decrease of mobility in the adsorbed two-dimensional phase, a given number of treatments have been suggested in which the adsorbed ions are localized at the apices of a hexagonal rigid lattice [65-68]. A lattice of this kind is open to question in the case of a liquid metal electrode at low or moderate coverage but, as mentioned later, can be a good model at appreciable surface coverage. However, if a regularly structured substrate strongly orders the adsorption, a model of localized adsorption is valid at any coverage. We now wish to indicate some very general expressions for such lattice adsorption without going into further details in the frame of this review.

Therefore, let us start from equation (126a). The quantity $\Lambda_\alpha{}^0(\rho_\alpha{}^A)$ is readily evaluated in the case of one particle–one site adsorption. Thus,

$$\Lambda_\alpha{}^0(\rho_\alpha{}^A) = \exp[ln(1 - \theta_c)]... \tag{147}$$

where θ_c is the degree of coverage; hence, $\theta_c = \rho_\alpha{}^A/N_s$ with N_s the number of sites per square centimeter. The first term in the integrand of equation (126b) represents, in the case of lattice adsorption, a linear coulombic interaction between α and its lattice neighbors. This assumption is quite valid and yields the following sum over the lattice points n and m:

$$-\frac{e_\alpha{}^2}{kT\epsilon_0} \theta_c \sum_{n,m} \gamma^A(s_{n,m}) \tag{148}$$

Under these conditions the microscopic potential becomes

$$\bar{\psi}_\alpha{}^M(ln) = \bar{\psi}(z_a) - \frac{q_a}{\epsilon_0} \int_{\mathrm{I}} ds_{12}\gamma^A(s_{12}) + \frac{e_0}{\epsilon_0} \theta_c \sum_{n,m} \gamma^A(s_{n,m}) \tag{149}$$

and

$$e_\alpha \bar{\psi}_\alpha{}^M(ln) = W_\alpha{}^{c(1)}$$

We consider now the two distinct cases where, respectively, $\epsilon_1 \gg \epsilon_0$

($\delta = 1$) and $\epsilon_1 \simeq \epsilon_0$ ($\delta_0 = 0$). For $\delta = 1$, it may be inferred from (137) that

$$\bar{\psi}_\alpha{}^M(1) = \frac{4\pi}{\epsilon_0} \left[\frac{h_d - z_a}{h_d} \right] [q^m h_d + q_a(h_d - z_a)] + \frac{e_\alpha}{\epsilon_0} \theta_c \sum_{n,m} \gamma^A(s_{n,m})$$

(150)

where the only effect of the upper limit of the integration in (149) is to cancel out the edge effects at $s = a$; thus, $\beta(h_d/a) = 0$. For a hexagonal lattice (150) has been solved by different authors [67,68]. As for the case $\delta = 0$, (149) yields

$$\bar{\psi}_\alpha{}^M(1) = \frac{4\pi}{\epsilon_0} \{q^m(h_d - z_a)\} + \frac{4\pi}{\epsilon_0} \{q_a(h_d - 2z_a)\}$$

$$+ \frac{e_\alpha}{\epsilon_0} M_{n,m} \theta_c \gamma^A(s_{n_0,m_0})$$

(151)

where $M_{m,n}$ is a Madelung constant depending on the lattice symmetry ($M_{m,n} = 11.034$ for a hexagonal array and 9.034 for a square array) [86]. Equations (150) and (151) must be modified in order to fit to the model of artificial lattice considered by Essin and Shikov [65]. For this case, θ_c must be dropped from the last term of our relations, and the lattice parameter is s_0.

The problem of thermal stability of an adsorbed regular array of charges has been treated recently by different authors [84,87,88]. In such treatments the central charge at (1) is allowed to move under thermal perturbation in the potential arising from all other charges fixed at the lattice apices and from all their electric images. As a rough criterion for lattice stability it has been suggested to assume the following condition [87] in the hexagonal case:

$$\frac{2}{\sqrt{3}} \frac{(\bar{s}_{12}^2)^{1/2}}{s_0} \leqslant 0.35$$

(152)

where $(\bar{s}_{12}^2)^{1/2}$ is the root-mean-square displacement of the considered ion relative to its position at (1); thus,

$$\bar{s}_{12}^2 = \frac{\displaystyle\int_{\bar{\psi}_\alpha^M(1)}^{\infty} s_{12}^2 \rho_\alpha(s_{12}) \exp - e_\alpha \bar{\psi}_\alpha{}^M(s_{12})/kT \, d\bar{\psi}_\alpha{}^M(s_{12})}{\displaystyle\int_{\bar{\psi}_\alpha^M(1)}^{\infty} \rho_\alpha(s_{12}) \exp - e_\alpha \bar{\psi}_\alpha{}^M(s_{12})/kT \, d\bar{\psi}_\alpha{}^M(s_{12})}$$

(153)

with $\rho_\alpha(s_{12})$ and $\bar{\psi}_\alpha^M(s_{12})$, respectively, the local density at zero micro-potential and the micropotential at a distance s_{12} from (1). On the basis of this treatment, MacDonald and Barlow [87] found that for monovalent ions, a single-image model nonorienting substrate, $q_a = -q^m$, and $z_a = 3$ Å, a hexagonal array with $s_0 = 15$ Å is stable up to a temperature T_0 given approximately by $1760/\epsilon_0°$K while one with $s_0 = 21$ Å is stable up to about $760/\epsilon_0°$K with ϵ_0, the dielectric constant in the inner layer.

Consequently, at a nonorienting substrate, the appropriateness of a model of individual particles, each vibrating in the constant energy well representing the average interaction with its neighbors, relies ultimately upon the existence of stable average positions for these charges, small oscillations or weakly coupled perturbations between neighbors, and sharp-shaped adsorption energy wells in order to keep coplanar location of adions irrespective of their strong lateral repulsion. Thus, in the case of mercury the treatment would actually amount to a choice of high coverages, relatively low kT, and to minimizing reflection in the OHP, which may substantially decrease the range of coulombic interactions as shown in Section 1, equation (7b). In the case of single imaging in the electrode, it has already been mentioned with respect to (124) that q_a larger than 10 μCb/cm^2 might be required before hexagonal order is established (cf. in this respect MacDonald and Barlow [87]). However, for the case of multiple reflections in the electrode and OHP, Levine et al. [88] were able to substantiate the argument that below $q_a < 15$–20 μCb/cm^2 the layer of adions should behave more like a two-dimensional dense gas than like a solid.

REFERENCES

1. Parsons, *Modern Aspects of Electrochem.*, Vol. 1, Bockris, ed., Butterworth, London (1954).
2. Smythe, *Static and Dynamic Electricity*, McGraw Hill Book Company, New York (1950).
3. Sneddon, *Fourier Transforms*, McGraw Hill Book Company, New York (1951).
4. Buff and Stillinger, Jr., *J. Chem. Phys.* **39**, 1911 (1963).
5. Krylov and Levich, *Z. Fiz. Khim.* **37**, 106 (1963).
6. Ershler, *Z. Fiz. Khim.* **20**, 679 (1946).
7. Prigogine, Mazur, and Defay, *J. Chim. Phys.* **50**, 146 (1953).
8. Sanfeld, Steinchen-Sanfeld, and Defay, *J. Chim. Phys.* **58**, 132 (1962).
9. Van de Berg, Ph.D. Dissertation, Free University of Brussels, Belgium (1964).
10. Sanfeld, Steinchen-Sanfeld, Hurwitz, and Defay, *J. Chim. Phys.* **58**, 139 (1962).

11. Sanfeld, *Bull. Classe Sci. Acad. Roy. Belg.* **3**, 339 (1964).
12. Hurwitz, Sanfeld, and Steinchen-Sanfeld, *Electrochim. Acta* **9**, 929 (1964).
13. Brown, *Am. J. Phys.* **19**, 290 (1951).
14. Owen, Miller, Milner, and Cogan, *J. Phys. Chem.* **65**, 2065 (1961).
15. Kirkwood, *J. Chem. Phys.* **7**, 911 (1939).
16. Lange, *Z. Elektrochem.* **56**, 94 (1952).
17. Defay and Mazur, *Bull. Soc. Chim. Belg.* **63**, 562 (1954).
18. de Groot and Tolhoek, *Proc. Koňink. Ned. Akad.* **B54**, 1 (1951).
19. Sanfeld, Thesis, Free University of Brussels, Belgium (1964).
20. Lange and Koenig, Handb. Experiment. Physik XII, 1933, Ak. Verlag Leipzig.
21. Gouy, *J. Phys.* **9**, 457 (1910).
22. Chapman, *Phil. Mag.* **25**, 475 (1913).
23. Helmholtz, *Ann. Physik.* **7**, 337 (1879).
24. Debye and Hückel, *Phys. Z.* **24**, 185 (1923).
25. Onsager, *Phys. Z.* **28**, 277 (1927); *Chem. Rev.* **13**, 73 (1933). Kirkwood, *J. Chem. Phys.* **2**, 767 (1934). Frank and Thomson, *J. Chem. Phys.* **31**, 1086 (1959).
26. Fowler and Guggenheim, *Statistical Thermodynamics*, Cambridge University Press, New York (1952).
27. Kirkwood and Poirier, *J. Phys. Chem.* **58**, 591 (1954).
28. Mayer, *J. Chem. Phys.* **18**, 1426 (1950).
29. Grimley and Mott, *Discussions Faraday Soc.* **43**, 3 (1947).
30. Eigen and Wicke, *Z. Elektrochem.* **55**, 354 (1951).
31. Eigen and Wicke, *Z. Elektrochem.* **56**, 836 (1952).
32. Freise, *Z. Elektrochem.* **56**, 836 (1952).
33. Hückel and Kraft, *Z. Phys. Chem. (N.F.)* **3**, 135 (1955).
34. Brodowsky and Strelow, *Z. Elektrochem.* **63**, 262 (1959).
35. Grahame, *J. Chem. Phys.* **18**, 903 (1950).
36. Conway, Bockris, and Ammar, *Trans. Faraday Soc.* **47**, 756 (1951).
37. Bikerman, *Phil. Mag.* **33**, 884 (1942).
38. Spaarnay, *Rec. Trav. Chim.* **77**, 382 (1958).
39. Grahame, *Chem. Rev.* **41**, 441 (1947).
40. Frumkin, *Advances in Electrochemistry*, Vols. 1 and 3, P. Delahay, ed., Interscience, New York (1961 and 1963).
41. Gierst, *These d'Agregation*, Free University of Brussels, Belgium (1958).
42. Hill, *Statistical Mechanics*, McGraw–Hill Book Company, New York (1956).
43. Friedman, *Ionic Solution Theory*, Interscience, New York (1962).
44. Stillinger and Buff, *J. Chem. Phys.* **37**, 1 (1962).
45. Stillinger and Kirkwood, *J. Chem. Phys.* **33**, 1282 (1960).
46. Hasted, Ritson, and Collie, *J. Chem. Phys.* **16**, 1 (1948).
47. Booth, *J. Chem. Phys.* **19**, 391 (1951).
48. Barlow and MacDonald, *J. Chem. Phys.* **36**, 3062 (1962).
49. Malsch, *Phys. Z.* **29**, 770 (1928); **30**, 837 (1929).
50. Hückel, *Phys. Z.* **26**, 93 (1925).
51. Sack, *Phys. Z.* **27**, 206 (1926); **28**, 199 (1927).
52. Verwey, *Rec. Trav. Chim.* **61**, 127 (1942).
53. Ackerman, *Discussions Faraday Soc.* **24**, 80 (1957); *Z. Phys. Chem.* **27**, 253 (1961).
54. Hasted and Roderick, *J. Chem. Phys.* **29**, 17 (1958).

55. Harned and Owen, *The Physical Chemistry of Electrolytic Solution*, 3rd ed., Reinhold, New York (1958).
56. Williams, *Proc. Phys. Soc.* **A66**, 372 (1953).
57. Hill, *J. Phys. Chem.* **548**, 61 (1957).
58. Stern, *Z. Elektrochem.* **30**, 508 (1924).
59. Parsons, *Advances in Electrochemistry*, Vol. 1, Delahay, ed., Interscience, New York (1961).
60. Delahay, *Double Layer and Electrode Kinetics*, Interscience, New York (1965).
61. Esin and Markov, *Acta Physicochim.* **10**, 353 (1939).
62. de Boer, Electron Emission and Adsorption Phenomena, Cambridge University Press, London (1935).
63. Langmuir, *J. Am. Chem. Soc.* **54**, 1252, 2798 (1932).
64. Frumkin, *Uspekhi Khim.* **4**, 938 (1935).
65. Esin and V. Shikhov, *Z. Fiz. Khim.* **17**, 236 (1943).
66. Iofa and Frumkin, *Z. Fiz. Khim.* **18**, 268 (1944).
67. Levich, Kiryanov, and Krylov, *Dokl. Akad. Nauk. Uz. SSR* **135**, 1425 (1960).
68. Barlow, Jr., and MacDonald, *J. Chem. Phys.* **40**, 1535 (1964); **43**, 2575 (1965).
69. Levine, Bell, and Calvert, *Can. J. Chem.* **40**, 518 (1962).
70. Grahame, *Z. Elektrochem.* **62**, 264 (1958).
71. Bockris, Devanathan, and Müller, *Proc. Roy. Soc. (London)* **A274**, 55 (1963).
72. Wroblowa, Kovac, and Bockris, *Trans. Faraday Soc.* **61**, 1523 (1965).
73. Barlow, Jr., and MacDonald, *J. Chem. Phys.* **39**, 412 (1963).
74. Rampolla, Miller, and Smyth, *J. Chem. Phys.* **30**, 566 (1959).
75. Sachs and Dexter, *J. Appl. Phys.* **21**, 1304 (1950).
76. Cutler and Gibbons, *Phys. Rev.* **111**, 394 (1958).
77. Helfand, Frisch, and Lebowitz, *J. Chem. Phys.* **34**, 1037 (1961).
78. Ree and Hoover, *J. Chem. Phys.* **40**, 939 (1964).
79. Parsons, *J. Electroanal. Chem.* **7**, 136 (1964).
80. Wertheim, *Phys. Rev. Letters* **8**, 321 (1963); *J. Math. Phys.* **5**, 643 (1964).
81. Throop and Bearman, *J. Chem. Phys.* **42**, 2408 (1965).
82. Krylov, *Electrochem. Acta* **9**, 1247 (1964).
83. Mignolet, *Bull. Soc. Roy. Sci. (Liege)* **23**, 422 (1954).
84. Levine, Mingins, and Bell, *Can. J. Chem.* **43**, 2834 (1965).
85. Hurwitz, in press.
86. Toppings, *Proc. Roy. Soc. (London)* **A114**, 67 (1927).
87. MacDonald and Barlow, Jr., *Can. J. Chem.* **43**, 2985 (1965).
88. Bell, Levine, and Mingins, IV International Congress on Surface Activity, Brussels, September 1964.
89. MacDonald, *J. Chem. Phys.* **22**, 1857 (1954).
90. Bell and Levine, *Chemical Physics of Ionic Solutions*, Conway and Barradas, eds., Wiley, New York (1966).

Index